Organic Chemistry
PEARLS OF WISDOM

THOMAS VALLOMBROSO
Hastings College

> **NOTE**
>
> The intent of Organic Chemistry Pearls of Wisdom is to serve as a study aid to improve performance on a standardized examination. Neither Boston Medical Publishing Corporation nor the editors warrant that the information in this text is complete or accurate. The reader is encouraged to verify each answer in several references.

Copyright © 2001 by Boston Medical Publishing Corporation, Lincoln, NE.

Printed in U.S.A.

All rights reserved, including the right of reproduction, in whole or in part, in any form.

The editors would like to extend thanks to Terri Lair for her excellent managing and editorial support.

Special thanks to Donald Schlegel, Professor of Sciences at College View Academy in Lincoln, Nebraska, for his review of this book.

This book was produced using Times New Roman, GillSans, Impact and Symbols fonts and computer based graphics with Macintosh® and PC computers. Figures were created with Chem Windows 3.0 and Paint Shop Pro 6.02

ISBN: 1-58409-016-2

DEDICATION

To my wife, Carmel, for her support and encouragement while I prepared this book, and to Milton Brown and Roy Olofson, two educators who demonstrated to me the joy of science and the joy of teaching.

Thomas Vallombroso

EDITOR

Thomas M. Vallombroso, Ph.D.
Associate Professor of Chemistry
Hastings College
Hastings, NE

WE APPRECIATE YOUR COMMENTS!

We appreciate your opinion and encourage you to send us any suggestions or recommendations. Please let us know if you discover any errors, or if there is any way we can make Pearls of Wisdom more helpful to you. We are also interested in recruiting new authors and editors. Please call, write, fax or e-mail. We look forward to hearing from you.

Return to:

Boston Medical Publishing Corporation
4780 Linden Street, Lincoln, NE, 68516

888-MBOARDS (626-2737)
402-484-6118
Fax: 402-484-6552
E-mail: bmp@emedicine.com
www.bmppearls.com

INTRODUCTION

Congratulations! Organic Chemistry Pearls of Wisdom will help you improve your knowledge base in Organic Chemistry. A few words are appropriate in discussing intent, format, limitations and use.

Since Pearls is primarily intended as a study aid, the text is written in a rapid-fire question/answer format. This way, readers receive immediate gratification. Moreover, misleading or confusing "foils" are not provided. This eliminates the risk of erroneously assimilating an incorrect piece of information that makes a big impression. Questions themselves often contain a "pearl" intended to reinforce the answer. Additional "hooks" may be attached to the answer in various forms, including mnemonics, visual imagery, repetition, and humor. Additional information, not requested in the question, may be included in the answer. Emphasis has been placed on distilling trivia and key facts that are easily overlooked, quickly forgotten and that somehow seem to be needed on MCAT, DAT, OAT and similar examinations.

Many questions have answers without explanations. This enhances ease of reading and rate of learning. Explanations often occur in a later question/answer. Upon reading an answer, the reader may think, "Hm, why is that?" or "Are you sure?" If this happens to you, go check! Truly assimilating these disparate facts into a framework of knowledge absolutely requires further reading of the surrounding concepts. Information learned in response to seeking an answer to a particular question is retained much better than information that is passively observed. Take advantage of this! Use Pearls with your preferred texts handy and open.

Pearls has limitations. We have found many conflicts between sources of information. We have tried to verify in several references the most accurate information. Some texts have internal discrepancies further confounding clarification.

Pearls risks accuracy by aggressively pruning complex concepts down to the simplest kernel—the dynamic knowledge base and practice of Organic Chemistry is not like that! Furthermore, new research and practice occasionally deviates from that which likely represents the right answer for test purposes. This text is designed to maximize your score on a test. Refer to your most current sources of information and mentors for further direction.

Pearls is designed to be used, not just read. It is an interactive text. Use a 3 x 5 card and cover the answers; attempt all questions. A study method we recommend is oral, group study, preferably over an extended meal or pitchers. The mechanics of this method are simple and no one ever appears stupid. One person holds Pearls, with answers covered and reads the question. Each person, including the reader, says "Check!" when he or she has an answer in mind. After everyone has "checked" in, someone states his/her answer. If this answer is correct, on to the next one; if not, another person says their answer or the answer can be read. Usually the person who "checks" in first receives the first shot at stating the answer. If this person is being a smarty-pants answer-hog, then others can take turns. Try it, it's almost fun!

Pearls is also designed to be re-used several times to allow, dare we use the word, memorization. Two check boxes are provided for any scheme of keeping track of questions answered correctly or incorrectly.

We welcome your comments, suggestions and criticism. Great effort has been made to verify these questions and answers. Some answers may not be the answer you would prefer. Most often this is attributable to variance between original sources. Please make us aware of any errors you find. We hope to make continuous improvements and would greatly appreciate any input with regard to format, organization, content, and presentation or about specific questions. We also are interested in recruiting new contributing authors and publishing new textbooks. Contact our manager at Boston Medical Publishing, Terri Lair, at (toll free) 1-888-MBOARDS. We look forward to hearing from you!

Study hard and good luck!

T.V.

TABLE OF CONTENTS

ALKANE PEARLS 9

ALKENE AND ALKYNE PEARLS 23

STEREOCHEMISTRY PEARLS 41

ALKYL HALIDE PEARLS 53

ALCOHOL PEARLS 71

ETHER AND EPOXIDE PEARLS 85

THIOL, SULFIDE AND DISULFIDE PEARLS 97

AROMATIC COMPOUND PEARLS 103

ALDEHYDE AND KETONE PEARLS 125

CARBOXYLIC ACID AND CARBOXYLIC ACID DERIVATIVE PEARLS 147

AMINE PEARLS 165

SPECTROMETRY PEARLS 175

BIBLIOGRAPHY 205

ALKANE PEARLS

Organic chemistry nowadays almost drives me mad. To me it appears like a primeval tropical forest full of the most remarkable things, a dreadful endless jungle into which one does not dare to enter for there seems to be no way out.
Friedrich Wohler (1835)

❏❏ **What are Hydrocarbons?**

Compounds containing only Carbon and Hydrogen.

❏❏ **What is a saturated hydrocarbon?**

A hydrocarbon containing only Carbon-Carbon single bonds and Carbon Hydrogen bonds.

❏❏ **What is an unsaturated hydrocarbon?**

A hydrocarbon containing at least one Carbon-Carbon multiple bond

❏❏ **What are the two main classes of hydrocarbons?**

Aliphatic hydrocarbons and aromatic hydrocarbons.

❏❏ **What are the major groups of aliphatic hydrocarbons?**

Alkanes, Alkenes, and Alkynes.

❏❏ **What are the two most typical reactions of Alkanes?**

Combustion and Halogenation.

❏❏ **What are the distinguishing structural features of an Alkane?**

Alkanes are made up of only Carbon and Hydrogen and have only C – C single bonds. Alkanes are saturated hydrocarbons.

❏❏ **What are the names of the first ten straight chain Alkanes.**

Methane, Ethane, Propane, Butane, Pentane, Hexane, Heptane, Octane, Nonane, and Decane.

❏❏ **What are the accepted names of the following sidechains?**

a) isopropyl, b) *tert*-butyl, c) *sec*-butyl, d) isobutyl

❏❏ **How do you recognize a *sec*-butyl sidechain?**

It has four Carbons in a straight chain and is attached to the parent chain through the second Carbon.

❏❏ **What feature does an isopropyl group, an isobutyl group, and an isopentyl group have in common?**

All sidechains whose name begins with "iso" end in a Y shaped branch.

❏❏ **In the common system of nomenclature, what do the prefixes "n", "neo" and "iso" mean?**

The prefix "n" refers to a straight chain, unbranched structure, "neo" is used to indicate a chain that terminates in a $(CH_3)_3C$ group, and "iso" designates an otherwise unbranched chain ending in a $(CH_3)_2CH$

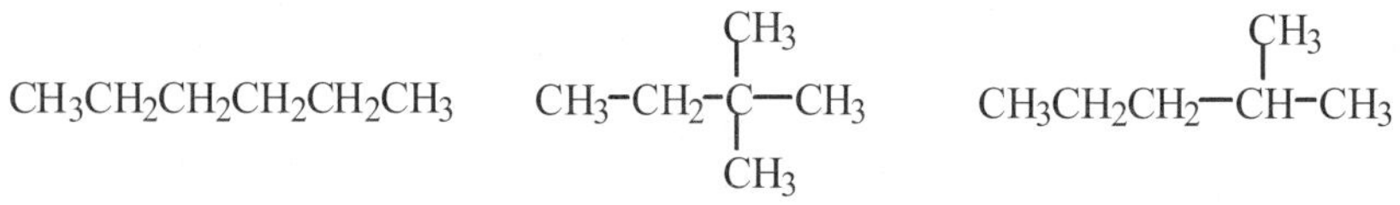

$CH_3CH_2CH_2CH_2CH_2CH_3$

$CH_3-CH_2-\underset{CH_3}{\overset{CH_3}{\underset{|}{\overset{|}{C}}}}-CH_3$

$CH_3CH_2CH_2-\overset{CH_3}{\overset{|}{CH}}-CH_3$

n-Hexane   Neohexane   Isohexane

❏❏ **What are the rules for naming a branched Alkane?**

1) Identify the longest continuous Carbon chain. This is the parent Alkane.
2) Identify and name substituent groups attached to the parent chain
3) Number the longest chain from the end that gives the lowest number to the substituent at the first branch
4) Write the parent Alkane at the end of the name and precede it by the names of the sidechains and their numerical location. The numerical locations are separated from the substituent names by a hyphen.

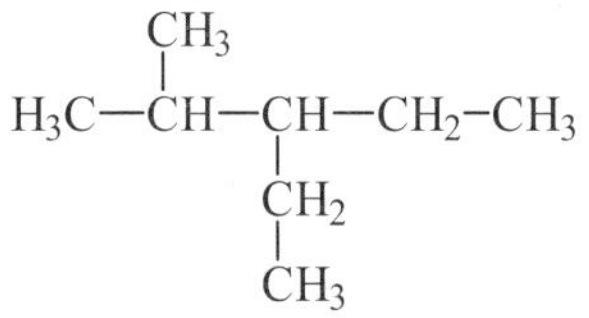

$H_3C-\overset{CH_3}{\overset{|}{CH}}-\underset{\underset{CH_3}{\overset{|}{CH_2}}}{\overset{|}{CH}}-CH_2-CH_3$ 3-Ethyl-2-methylpentane

❏❏ **In an Alkane containing multiple substituents, how does one order the substituents in the name?**

Alphabetically, not considering hyphenated prefixes such as *sec-* and *tert-* in the alphabetization.

❏❏ **What are constitutional isomers?**

Compounds that have the same molecular formula, but differ in the order in which the atoms are connected.

❏❏ **How do the physical properties of constitutional isomers compare?**

The physical properties are different. For hydrocarbon isomers, the properties may be similar, but if atoms other than Carbon and Hydrogen are present, the properties can differ quite a bit.

❏❏ **How many constitutional isomers are there of the formula C_6H_{14}?**

Five:

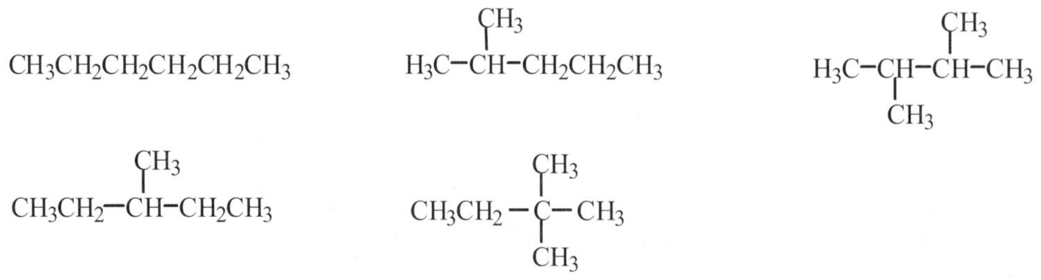

$CH_3CH_2CH_2CH_2CH_2CH_3$ $H_3C-\overset{CH_3}{\overset{|}{CH}}-CH_2CH_2CH_3$ $H_3C-\underset{\underset{CH_3}{\overset{|}{}}}{\overset{CH_3}{\overset{|}{CH}}}-CH-CH_3$

$CH_3CH_2-\overset{CH_3}{\overset{|}{CH}}-CH_2CH_3$ $CH_3CH_2-\underset{CH_3}{\overset{CH_3}{\underset{|}{\overset{|}{C}}}}-CH_3$

❑❑ **What is the best way to determine the number of constitutional isomers of a particular formula?**

There is no easy way to determine the number of isomers. The best way is to draw all of the possibilities, working systematically from the unbranched chain, then shortening it while adding branches.

❑❑ **What are Cycloalkanes?**

An Alkane is attached end to end forming a ring.

❑❑ **How are simple Cycloalkanes named?**

By taking the name of the straight chain Alkane with the same number of Carbons as the ring, and adding the prefix "cyclo".

❑❑ **What is a fused ring system?**

A molecule in which two or more rings share at least one side.

❑❑ **What is a spiro compound?**

A compound in which two rings share a single Carbon.

❑❑ **What are the IUPAC standard names of the following compounds?**

(a) 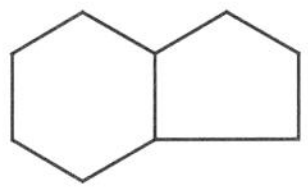 Bicyclo [4.3.0] nonane, (b) Spiro [2.4] heptane, (c) 2,3,4-Trimethylbicyclo [4.4.0] decane, (d) Bicyclo [2.2.1] heptane

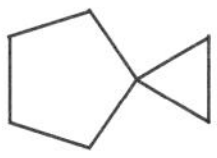

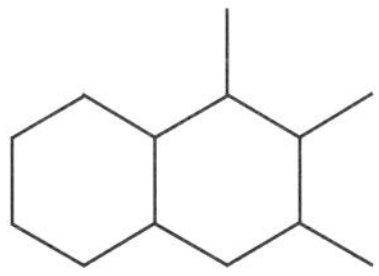

❑❑ **What are conformations?**

Representations of the same molecule that differ only in their position of rotation around a Carbon-Carbon single bond.

❑❑ **What three types of representations are commonly used to depict the conformations of ethane?**

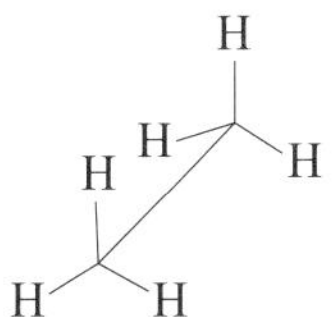

Sawhorse 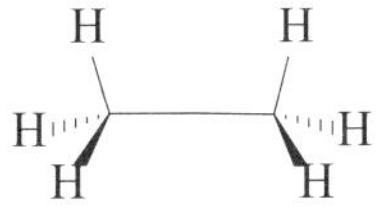   Wedge and Dash 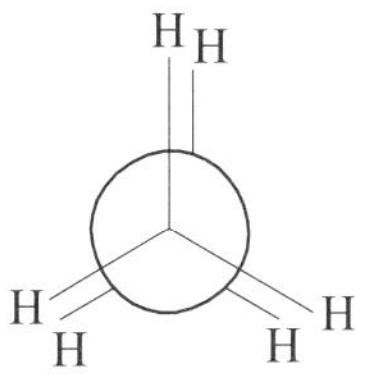   Newman Projection

❑❑ **Using Newman projections, draw the staggered and eclipsed conformations of Ethane.**

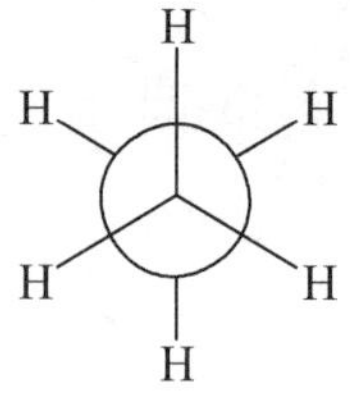

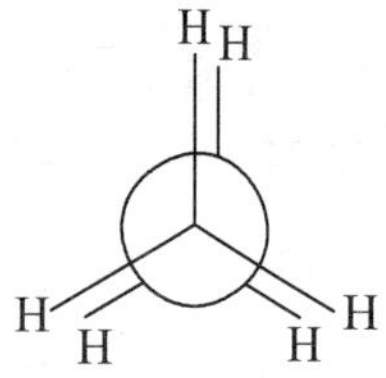

Staggered   Eclipsed

❑❑ **What do we call the destabilization caused by the eclipsing of bonds on adjacent atoms?**

Torsional Strain.

❑❑ **What is the energy difference between the eclipsed and staggered forms of Ethane?**

12 kJ/mol or 2.9 kcal/mol.

❑❑ **In a Newman projection, what do we call two groups that are 180° apart?**

Anti.

❑❑ **What do we call two groups that are 60° apart?**

Gauche.

❑❑ **What is Steric Hindrance?**

The destabilization of a molecule caused when two of its atoms get too close together. Also known as steric strain.

❑❑ **Is it possible to convert *cis*-1, 2-Dimethylcyclohexane to *trans*-1, 2-Dimethylcyclohexane by changing the conformation?**

No. Such a transformation can only be accomplished by breaking and re-forming bonds.

❑❑ **Draw the chair conformation of Cyclohexane.**

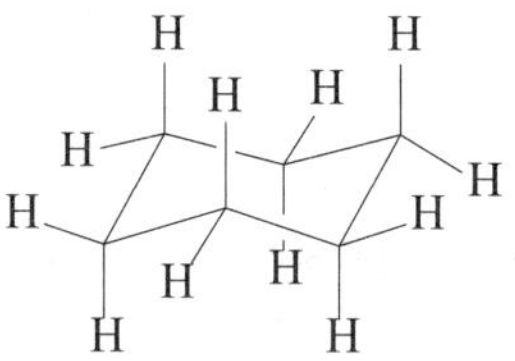

❑❑ **How many unique environments are there for the Hydrogens in the Cyclohexane chair conformation?**

Two.

❑❑ **In the Cyclohexane chair conformation, what is meant by the term Axial?**

Axial refers to the six substituent positions whose bonds are oriented up and down.

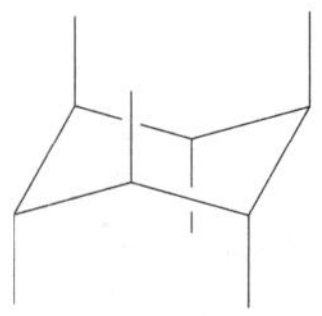

❑❑ **What is meant by Equatorial?**

These are the positions whose bonds are oriented out from the chair, around its "equator."

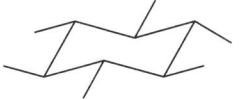

❑❑ **Is there a significant difference between these two positions?**

Yes, the axial position is more crowded.

❑❑ **What are 1,3-diaxial interactions?**

They are steric interactions between the substituents of the cyclohexane chair that reside in the axial positions and share the same face of the ring.

❑❑ **What is a skew conformation?**

Any conformation between staggered and eclipsed.

❑❑ **What forces determine the physical states of Alkanes?**

As nonpolar molecules, the physical states of Alkanes are determined by the van der Waals forces between the molecules. These forces increase with increasing molecular size and decrease with increased branching.

☐☐ **Carbons in Alkanes are often classified into four types. What are these types?**

Primary (1°), Secondary (2°), Tertiary (3°), and Quaternary (4°). A Primary Carbon has only one other Carbon attached, a Secondary Carbon is joined to two other Carbons, a Tertiary Carbon is bonded to three Carbons, and a Quaternary Carbon has all four of it's valence positions filled by Carbons.

☐☐ **Draw the two chair conformations of *trans*-1,2-dimethylcyclohexane. Which of these two conformations is the most stable?**

Most Stable - Both methyls equitorial

The conformation that places both methyls equitorial is the most stable because this conformation reduces 1,3-diaxial interactions between the methyl groups and the Hydrogens of the ring.

☐☐ **Draw the two chair conformations of *cis*-1-bromo-4-*tert*-butylcyclohexane. Which conformation is the most stable?**

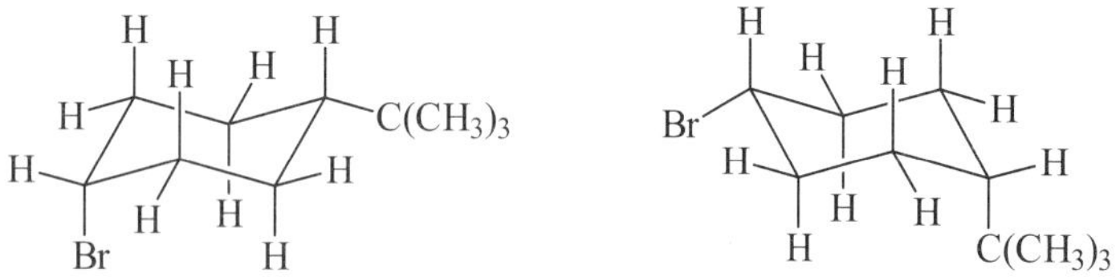

Most Stable - Bulkier *tert*-butyl group is equitorial.

The most stable conformation is the one that places the bulky tert-butyl group in the equitorial position. Large groups tend to prefer the equitorial position in order to reduce steric strain.

☐☐ **What are the structures of the following Alkanes?**

a) 2,2-Dimethylbicyclo[3.2.1]octane
b) 2-Ethylbicyclo[2.2.2]octane
c) Bicyclo[3.3.1]nonane
d) 2-Cyclobutylbicylco[2.1.1]hexane

a)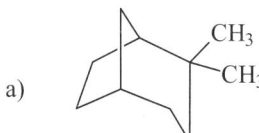

b)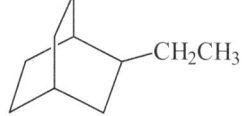

c)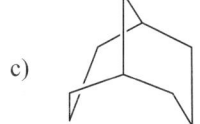

d)

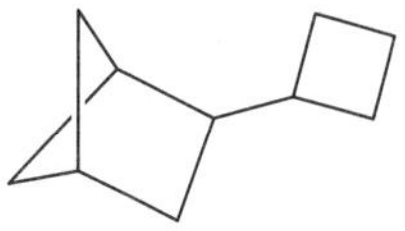

❏❏ **Match the following terms.**

1) **Dihedral Angle**
2) **IUPAC**
3) **C_nH_{2n+2}**
4) **Flagpole Interaction**
5) **Dispersion Forces**
6) **Bicycloalkane**

a) **Bridgehead Atom**
b) **Steric Strain**
c) **Induced Dipole**
d) **Conformation**
e) **Nomenclature**
f) **Alkanes**

(1) d, (2) e, (3) f, (4) b, (5) c, (6) a

❏❏ **What are the products of complete combustion of an Alkane?**

The products of complete combustion are Carbon dioxide and Water.

❏❏ **What dangerous product is produced by incomplete combustion of Alkanes?**

Carbon monoxide.

❏❏ **What product(s) would be obtained from the chlorination of Methane with one equivalent of Chlorine?**

Chlorination with equimolar amounts of reactants would produce a mixture of all possible products: CH_3Cl (Chloromethane), CH_2Cl_2 (Dichloromethane or Methylene Chloride), $CHCl_3$ (Trichloromethane or Chloroform), and CCl_4 (Tetrachloromethane or Carbon Tetrachloride).

❏❏ **How could you obtain Chloromethane as the primary chlorination product of Methane?**

Use an excess of methane.

❏❏ **Aside from chloromethanes, what is the other byproduct of the chlorination of methane?**

HCl.

❏❏ **What are the three general steps of the mechanism for the halogenation of Alkanes?**

Initiation, propagation and termination.

❏❏ **What does it take to initiate a chlorination or bromination reaction?**

Heat or light.

❏❏ **What phrase could be used to describe the mechanism of Alkane halogenation?**

Radical chain reaction.

❏❏ **Arrange the following molecules in order of increasing boiling point.**

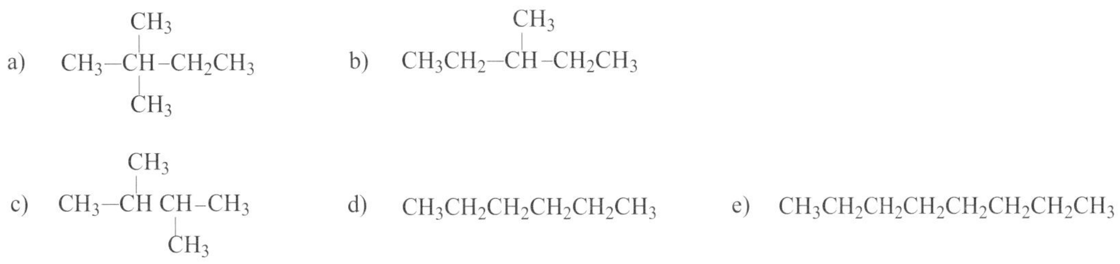

a) $CH_3-CH(CH_3)-CH_2CH_3$ with additional CH_3 (2,2-dimethyl substitution)

b) $CH_3CH_2-CH(CH_3)-CH_2CH_3$

c) $CH_3-CH(CH_3)\,CH(CH_3)-CH_3$

d) $CH_3CH_2CH_2CH_2CH_2CH_3$

e) $CH_3CH_2CH_2CH_2CH_2CH_2CH_2CH_3$

In order of increasing boiling point, a, c, b, d, e. Compounds a-d are constitutional isomers. Only the degree of branching will determine their relative boiling points, with the least branched having the highest bp. Compound e has a larger number of Carbons than the others, and hence the highest boiling point.

❏❏ **Draw Newman projections for all the staggered and eclipsed conformations of butane formed by rotation from 0° to 360° around the bond between Carbons 2 and 3.**

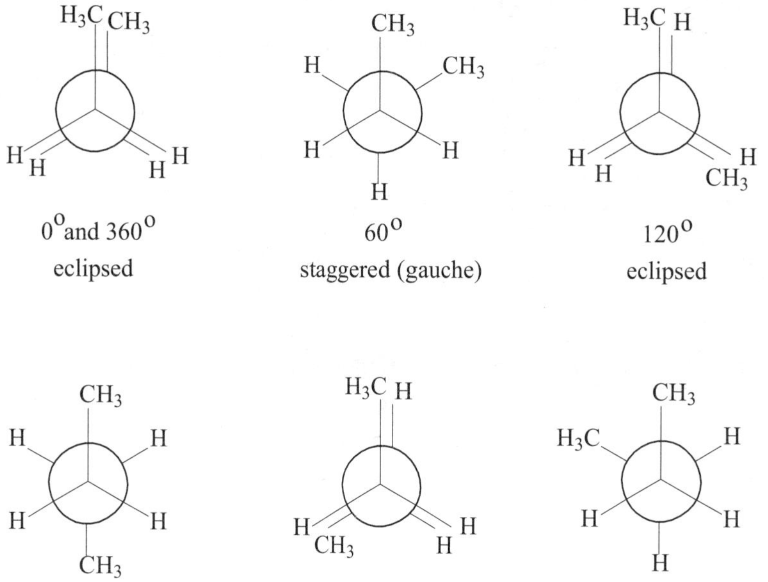

0° and 360° eclipsed

60° staggered (gauche)

120° eclipsed

180° staggered (anti)

240° eclipsed

300° staggered (gauche)

❏❏ **Aside from the chair conformations, what are the other most stable conformations of a Cyclohexane ring?**

There are two "Twist boat" conformations that are about 5.5 kcal/mol higher in energy than the chair conformations.

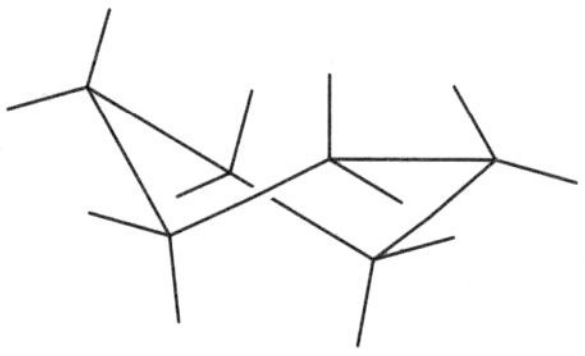

❏❏ **In the bromination of 2-Methylpropane, what is the order of reactivity of the two types of Hydrogens?**

The tertiary proton is much more reactive than the primary protons. In monobromination, the yield of 2-Bromo-2-methylpropane is greater than 99%, and the yield of 1-Bromo-2-methylpropane less than 1 %. In both bromination and chlorination, the order of reactivity of protons is $3° > 2° > 1°$.

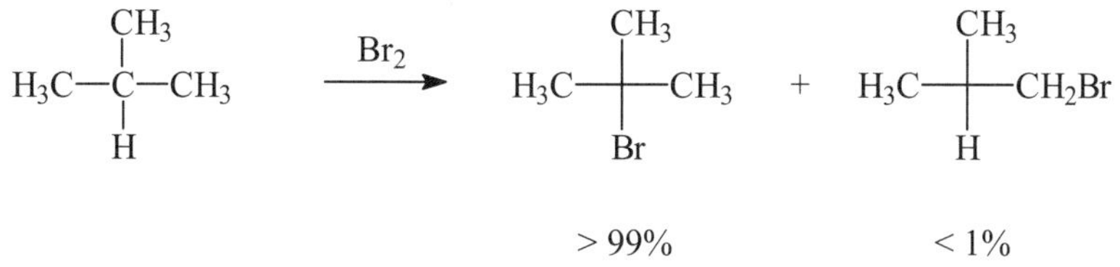

$$H_3C-\underset{\underset{H}{|}}{\overset{\overset{CH_3}{|}}{C}}-CH_3 \xrightarrow{Br_2} H_3C-\underset{\underset{Br}{|}}{\overset{\overset{CH_3}{|}}{C}}-CH_3 + H_3C-\underset{\underset{H}{|}}{\overset{\overset{CH_3}{|}}{C}}-CH_2Br$$

$$> 99\% \qquad\qquad < 1\%$$

❏❏ **What kind of bonding is found in Alkanes?.**

In Alkanes, the Carbons are all sp^3 hybridized. The bonds between Carbons are σ bonds formed by the overlap between sp^3 orbitals on adjacent Carbons. The C-H bonds are σ bonds resulting from overlap of the Carbon sp^3 orbitals and the 1s orbitals of the Hydrogens.

❏❏ **How acidic are Alkanes?**

Alkanes are not very acidic at all. The pK_a of an Alkane Hydrogen is usually about 50.

❏❏ **Match the following terms.**

1) Spiro Compound
2) Methylene
3) Angle Strain
4) Envelope
5) Cracking
6) Substituent

a) Preferred conformation of cyclopentane
b) A method of converting Alkanes to unsaturated hydrocarbons
c) -CH$_2$- group
d) Two rings share a single Carbon
e) A side chain or group
f) Found in small rings.

(1) d, (2) c, (3) f, (4) a, (5) b, (6) e

☐☐ What does the potential energy diagram for rotation around the C-C bond of ethane look like?

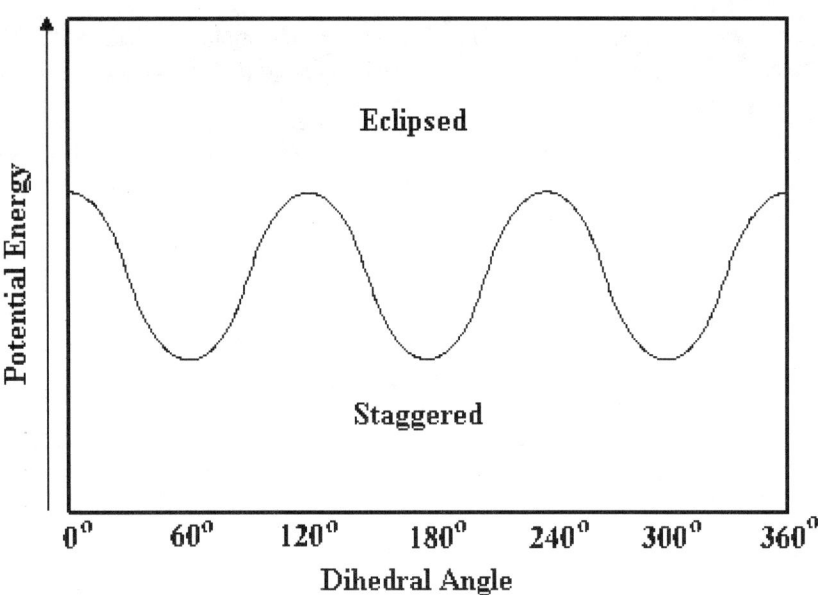

☐☐ What are the IUPAC names for the following compounds?

(a) *trans*-1,3-Dimethylcyclopentane, (b) *trans*-1,3-Dimethyl-2,2-diethylcyclohexane, (c) 1,1-Diethyl-3,3-dimethylcyclobutane, (d) 2-Methyl-4-ethylheptane

☐☐ **How many types of equivalent Hydrogens are in each of the following molecules?**

a) 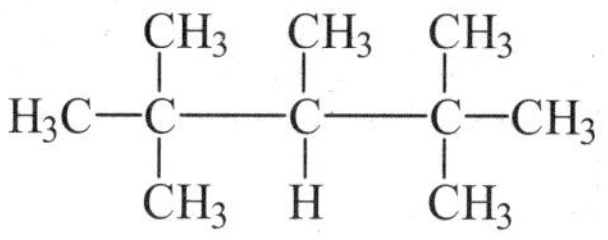
```
    CH3  CH3  CH3
     |    |    |
H3C—C————C————C—CH3
     |    |    |
    CH3   H   CH3
```

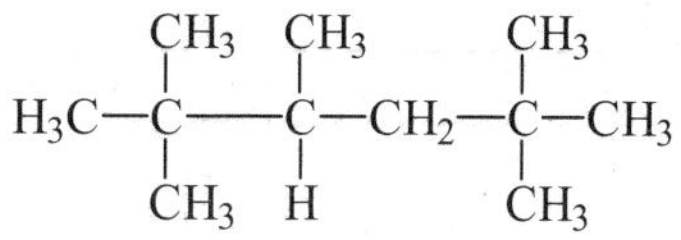

b)
```
    CH3  CH3        CH3
     |    |          |
H3C—C————C————CH2————C—CH3
     |    |          |
    CH3   H         CH3
```

c)
```
H3C—CH2
    |
    CH2—CH2—CH2—CH3
```

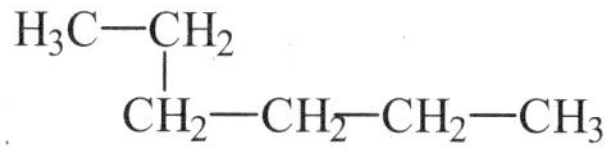

d)

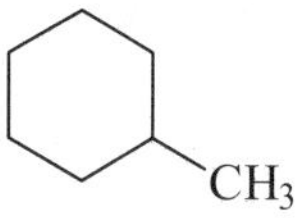

(a) 3, (b) 5, (c) 3, (d) 5.

☐☐ **In relation to Cycloalkanes, what is meant by the term cis/trans isomerism?**

Two substituents attached to a ring may be on the same side (face) of the ring (Cis), or they may be on opposite sides of the ring (Trans).

☐☐ **How many geometric isomers are possible for each of the of the following compounds?**

a)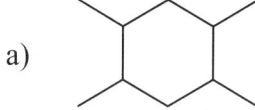

b)

c)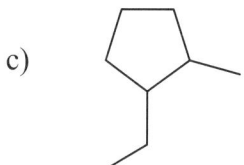

d)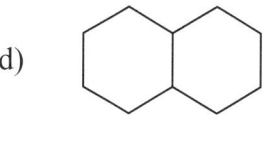

(a) 5, (b) 1, (c) 2, (d) 2.

☐☐ **Match each of the following phrases with the chemical structure it describes.**

1) Highly Strained
2) Flagpole Interactions
3) Locks into one conformation
4) Has bond angles of about 88 degrees

a)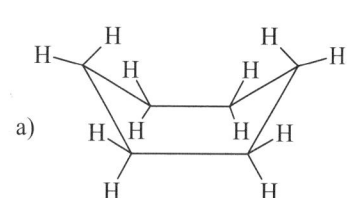

c)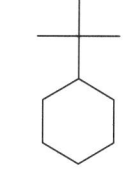

b)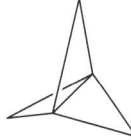

d) ☐

(1) b, (2) a, (3) c, (4) d.

❏❏ **What is the typical bond angle in an Alkane?**

109.5 degrees. Alkane Carbons are sp³ hybridized and therefore tetrahedral.

❏❏ **Why is Cyclopropane such a strained molecule?**

The three membered ring makes the bond angle 60 degrees, significantly lower than the 109.5 degrees preferred by sp³ hybridized Carbons. In addition, since the Cyclopropane ring must be flat, all of the Hydrogens are eclipsed. This contributes torsional strain.

❏❏ Draw the structures of all of the possible monochlorination products of the following molecules. Indicate which of the products is likely to be formed in the highest yield.

a) 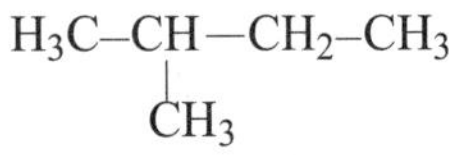 H₃C–CH(CH₃)–CH₂–CH₃

b) 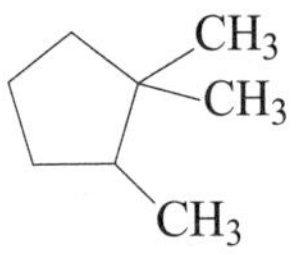 cyclopentane with CH₃, CH₃, CH₃ substituents

c) 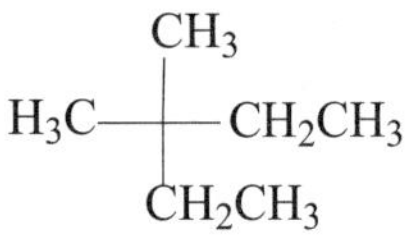 H₃C–C(CH₃)(CH₂CH₃)–CH₂CH₃

a) H₂C(Cl)–CH(CH₃)–CH₂–CH₃   [H₃C–C(Cl)(CH₃)–CH₂–CH₃]   H₃C–CH(CH₃)–CH(Cl)–CH₃   H₃C–CH(CH₃)–CH₂–CH₂Cl

b) cyclopentane products — the boxed (highest yield) product is the one with Cl on the quaternary-adjacent tertiary carbon bearing CH₃.

c) CH₂Cl–C(CH₃)(CH₂CH₃)–CH₂CH₃   [H₃C–C(CH₃)(Cl)–CH(CH₃)–CH₂CH₃ style — boxed: H₃C–C(CH₃)(CH₂CH₃)–CHCl–CH₃]   H₃C–C(CH₃)(CH₂CH₃)–CH₂–CH₂Cl

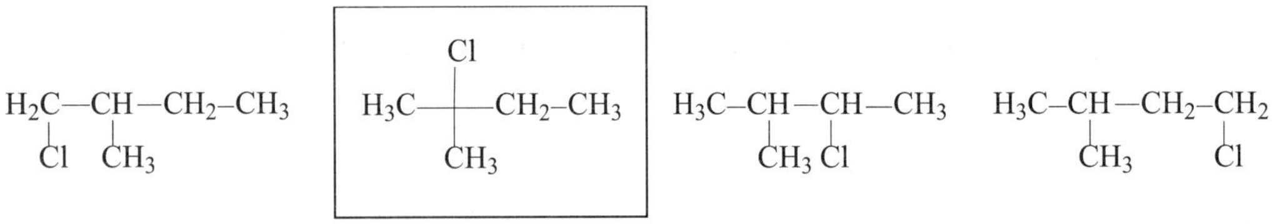

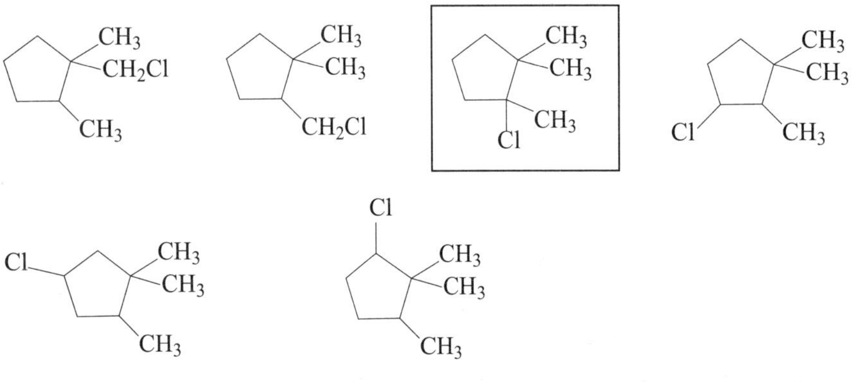

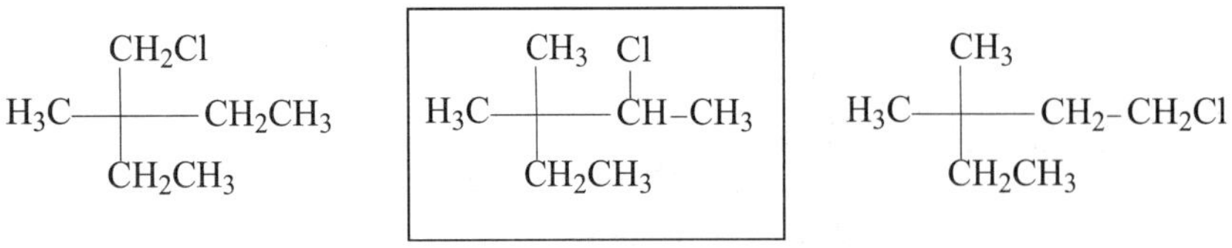

❏❏ What is meant by "degree of unsaturation"?

It is the number of rings or double bonds present in an unknown. For hydrocarbons, it may be determined by comparison of the unknown's molecular formula with the formula of a saturated Alkane ($C_n H_{2n+2}$). For each pair of Hydrogens missing from the formula of the unknown, one degree of unsaturation is present. For example, an unknown of the formula C_6H_8 has three degrees of unsaturation, since a saturated Alkane of the same number of Carbons would have 14 Hydrogens. $(14 - 8)/2 = 3$ degrees of unsaturation.

❏❏ Draw all of the possible saturated constitutional isomers of the formula C_6H_{12}. Ignore geometric isomers.

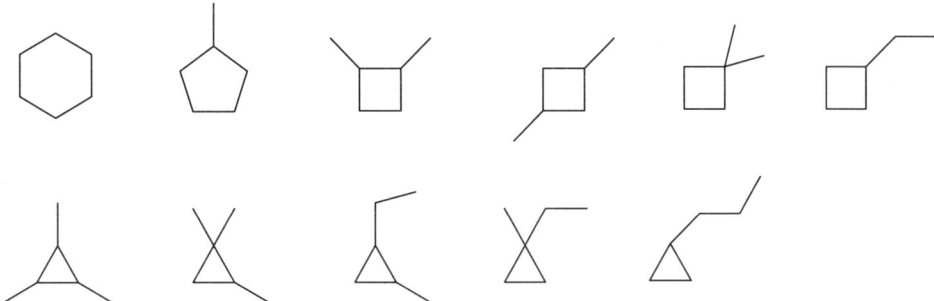

ALKENE AND ALKYNE PEARLS

Every attempt to employ mathematical methods in the study of chemical questions must be considered profoundly irrational and contrary to the spirit of chemistry. ... If mathematical analysis should ever hold a prominent place in chemistry -- an aberration which is happily almost impossible -- it would occasion a rapid and widespread degeneration of that science.
Auguste Comte, Philosophie Positive, 1838

❏❏ **What is an Alkene?**

An Alkene is a hydrocarbon with at least one Carbon-Carbon double bond.

❏❏ **What word was formerly used as the name of the class of compounds now known as Alkenes?**

Olefins.

❏❏ **What is an Alkyne?**

A hydrocarbon that has at least one Carbon-Carbon triple bond.

❏❏ **What is the difference between an internal and a terminal Alkyne?**

An internal Alkyne has the triple bond in the middle of a Carbon chain, and a terminal Alkyne has the triple bond at the end of a chain, with a C-H bond at one end.

❏❏ **What is the simplest Alkene called?**

Ethylene (common name) or ethene (IUPAC).

❏❏ **What is the simplest Alkyne called?**

Acetylene (common name) or ethyne (IUPAC).

❏❏ **What is the hybridization of the doubly bonded Carbon in an Alkene?**

The hybridization is sp^2.

❏❏ **What is the hybridization of the triply bonded Carbon in a Alkyne?**

The hybridization is sp^1.

❏❏ **What important feature do Alkenes and Alkynes have in common?**

Both of these classes of compounds have C-C π bonds formed through the overlap of unhybridized 2p orbitals.

❏❏ **Name the following compunds using IUPAC standard nomenclature.**

a) 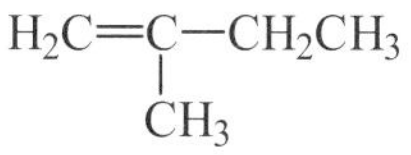 $H_2C=C(CH_3)-CH_2CH_3$ b) 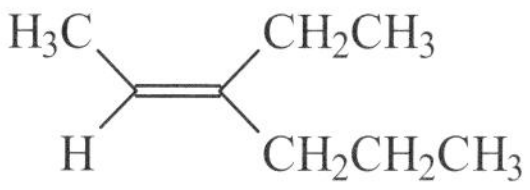 $(H_3C)(H)C=C(CH_2CH_3)(CH_2CH_2CH_3)$ c)

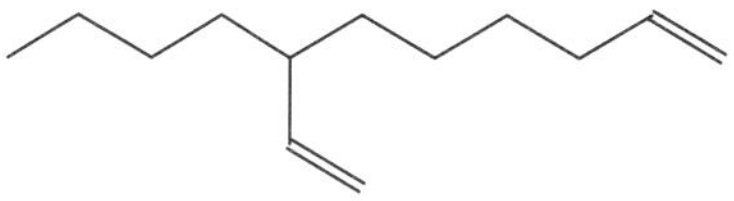

a) 2-Methyl-1-butene, b) 3-Ethyl-2-hexene, c) 3-Butyl-1, 8-nonadiene.

❑❑ **What is the most characteristic type of Alkene reaction?**

Electrophilic addition to the double bond.

❑❑ **What is the difference between a σ bond and a π bond?**

A σ bond has its highest electron density along the axis between the two bonded atoms. A π bond has its highest electron density above and below the internuclear axis.

　　σ bond　　　π bond

❑❑ **Which is stronger, a Carbon-Carbon single bond or a Carbon-Carbon double bond?**

The double bond is stronger because it consists of both a π bond and a σ bond. The single bond is just a σ bond.

❑❑ **Match the following terms.**

1) Hydration　　　　　　a) -CH₂=CH₂
2) Vinyl Group　　　　　 b) Designates Configuration
3) Halohydrin　　　　　　c) Addition of Water
4) Terpene　　　　　　　 d) Water + Halogen
5) E-Z　　　　　　　　　 e) Naturally Occurring Alkene

(1) c, (2) a, (3) d, (4) e, (5) b.

❑❑ **IUPAC has accepted the common names of three alkenyl groups. What are they?**

Vinyl, Allyl, and Isopropenyl.

❑❑ **What are the structures of these alkenyl groups?**

CH₂=CH—　　　CH₂=CH—CH₂—　　　CH₂=C—
　　　　　　　　　　　　　　　　　　　　|
　　　　　　　　　　　　　　　　　　　　CH₃

　Vinyl　　　　　　Allyl　　　　　　Isopropenyl

❑❑ **Why is it not necessary to give a number to indicate the position of the double bond in an unsubstituted Cycloalkene?**

It is understood that the double bond connects Carbon-1 and Carbon-2.

❑❑ **How do you number the ring of a substituted Cycloalkene?**

You begin with the double bond, proceed through it, and continue around the ring. The direction of numbering is chosen to give the lowest possible numbers to the substituents.

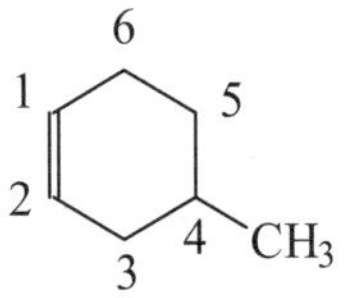

　Correct 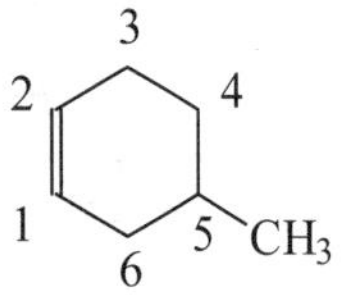　　　　　Incorrect

❏❏ **What is meant by the term Regioselective?**

A regioselective reaction is one in which one direction of bond forming or bond breaking occurs in preference to all others.

❏❏ **In Alkenes, what is meant by the terms Cis and Trans?**

These refer to the geometry around the double bond. In a cis Alkene, a pair of matching substituents on adjacent Carbons appears on the same side of the double bond. In a trans Alkene, the substituents are on opposite sides.

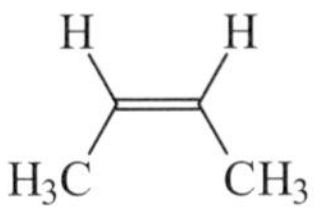

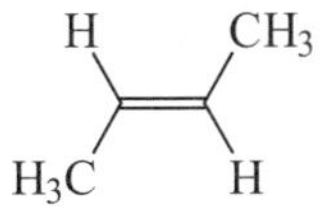

❏❏ **What is the Cahn-Ingold-Prelog system?**

It is a series of sequence rules for assigning priorities to substituent groups, such as those attached to a double bond.

❏❏ **What is meant by the designations E and Z?**

This is another way of specifying double bond geometry. The four groups around the double bond are assigned priorities according to the Cahn-Ingold-Prelog system. If the higher priority groups on each Carbon are on the same side of the double bond, the designation is Z, if they are on opposite sides, the designation is E. (a good way of remembering this is Z = "Zame Zide")

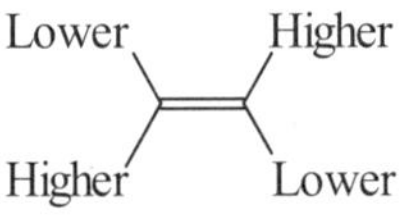

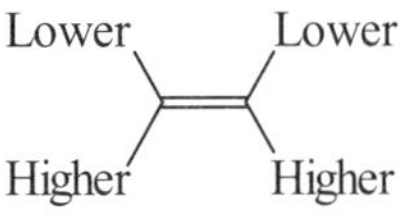

❏❏ **What are the configurations (if any) around the double bonds of the following Alkenes?**

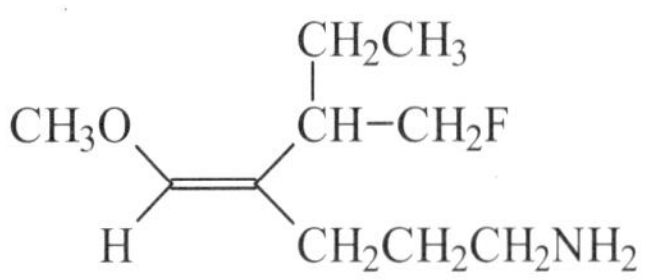

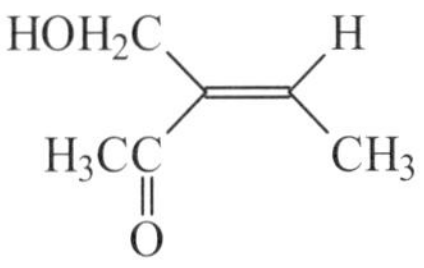

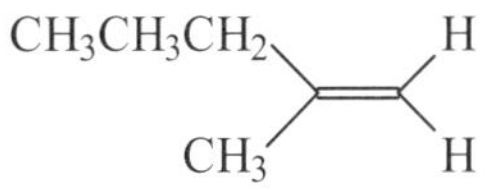

(a) 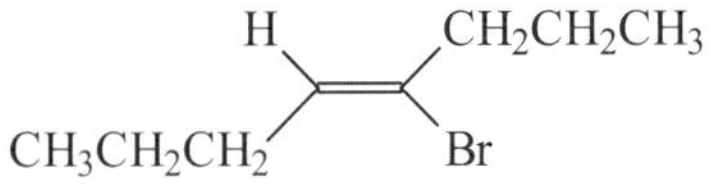 Z, (b) Z, (c) Z, (d) No configuration

❏❏ **The stereoselectivity of the Hydroboration/Oxidation reaction of Alkenes may best be described by what word?**

Syn.

❏❏ **What features do the Mercuration and Bromination reactions have in common?**

Both are believed to go through a three-membered ring intermediate, and both result in anti addition.

❏❏ **Why are addition reactions that go through a carbocation intermediate not stereoselective?**

Because the carbocation formed is sp^2 hybridized, and therefore flat. Addition of a nucleophile may occur from either side with equal ease, thus yielding no stereoselectivity.

❏❏ **What drawbacks do carbocation intermediates have?**

Carbocations are prone to rearrangement, and therefore can give "unexpected" products and mixtures.

❏❏ **What is Markovnikov's Law?**

In the addition of HX to an Alkene, the Hydrogen will end up on the Carbon that already has the largest number of Hydrogens. (The rich get richer. Those that have, get.)

❏❏ **Name three ways that "water" can be added to an Alkene. What are the practical features of these methods?**

<u>Acid catalysed addition of H_2O</u>: Harsh conditions. Markovnikov addition, but carbocation is prone to rearrangement. No stereoselectivity.

<u>Oxymercuration/Reduction</u>: Toxic reagents. Markovnikov addition, anti stereochemistry. No rearrangement because there is no carbocation intermediate.

<u>Hydroboration/Oxidation</u>: Anti-Markovnikov addition, syn stereochemistry. No rearrangement because there is no carbocation intermediate.

❏❏ **What is the most typical way of reducing an Alkene?**

Reaction with Hydrogen in the presence of a transition metal catalyst (platinum, palladium, nickel, etc), usually at elevated pressures.

❏❏ **What other names are used for this process?**

Catalytic reduction or catalytic hydrogenation.

❏❏ **Is catalytic hydrogenation exothermic or endothermic?**

Exothermic (heat of hydrogenation for Ethylene = -32.8 kcal/mol).

❏❏ **What is the stereochemistry of catalytic hydrogenation?**

Addition of Hydrogen is primarily Syn.

☐☐ **What is the mechanism for catalytic hydrogenation?**

Transition metals absorb Hydrogen molecules onto their surfaces by forming metal-hydrogen sigma bonds, which effectively weakens the H-H bonds. Alkenes are also absorbed onto the surface, with formation of Carbon-metal sigma bonds and cleavage of the C-C pi bond. A Hydrogen atom is then transferred to the Alkene from the metal surface, giving an intermediate in which the former Alkene is still attached to the metal. In a second step, another Hydrogen atom is transferred from the same side and the fully reduced Alkene is released.

☐☐ **The following heats of hydrogenation have been determined for the two isomeric forms of 2-Butene. Based on this data, explain which is the more stable isomer and why you believe it to be so.**

cis-2-Butene
-28.6 kcal/mol

trans-2-Butene
-27.6 kcal/mol

The *trans* isomer is the more stable structure. Both of the 2-Butenes are converted to the same product (n-Butane) by hydrogenation. The *cis* isomer releases 28.6 kcal in this transformation, indicating that it is 28.6 kcal/mol higher in energy than Butane. The *trans* isomer only releases 27.6 kcal/mol, indicating that it is only 27.6 kcal higher than Butane, and therefore 1 kcal/mol more stable than the *cis* isomer.

☐☐ **Of all the hydrocarbons, the Alkyne family possesses one unique feature. What is it?**

Members of one class of the Alkynes, the terminal Alkynes, are acidic. The proton on the terminal Carbon may be removed using an Amide base.

☐☐ **What is the pK_a of the proton on a terminal Alkyne?**

The pK_a is about 25.

☐☐ **Why are internal Alkynes not acidic?**

There are no protons on the sp hybridized Carbons.

☐☐ **What reagents may be used to convert a terminal Alkyne into an Aldehyde?**

Aldehydes are the result of treatment of a terminal Alkyne with a hindered dialkylborane such as disiamylborane {$(sia)_2BH$}, followed by reaction with H_2O_2/NaOH.

☐☐ **Which bond is stronger, a double bond or a triple bond? Why?**

A triple bond is stronger for two reasons: (1) a triple bond consists of a sigma bond and two pi bonds, whereas a double bond consists of only a sigma bond and a single pi bond., (2) the C-C bond distance in Alkynes is shorter due to the Carbon sp hybridization. This allows for better orbital overlap and stronger bonds.

☐☐ **Why are Alkenes flat?**

The pi bond in Alkenes is the result of overlap between two unhybridized p orbitals residing on sp^2 hybridized Carbons. These p orbitals are arranged perpendicular to the sp2 orbitals that participate in the other bonds. The p orbitals must be aligned side by side for the pi bond to exist, rendering the molecule flat.

☐☐ **What is the pK_a of a proton attached to a doubly bonded Carbon?**

About 45.

☐☐ **What would be the expected primary organic product of each of the following reactions?**

a) $CH_3CH_2-C\equiv CH$ + HBr (excess) ⟶

b)  [cyclopentene with H, H] + Br_2 ⟶

c) $CH_3C\equiv CCH_2CH_2CH_3$ + H_2 —Lindlar Catalyst→

d) 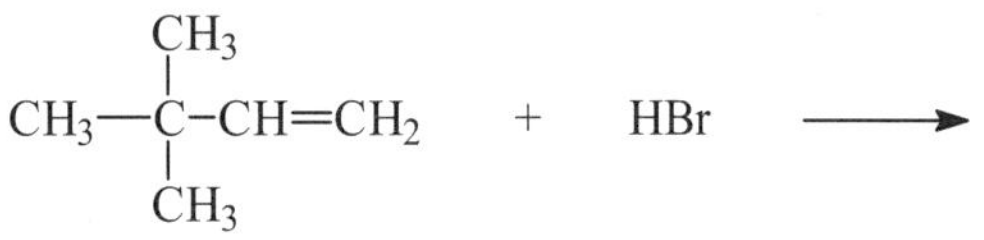 $CH_3-\underset{CH_3}{\overset{CH_3}{C}}-CH=CH_2$ + HBr ⟶

a) $CH_3CH_2-\underset{Br}{\overset{Br}{C}}-CH_2$

b) [cyclopentane with Br, H, H, Br] + [cyclopentane with H, Br, Br, H]

c) [alkene: CH_3 and $CH_2CH_2CH_3$ on same side, H and H on other]

d) $CH_3-\underset{Br}{\overset{CH_3}{C}}-\underset{CH_3}{CH}-CH_3$

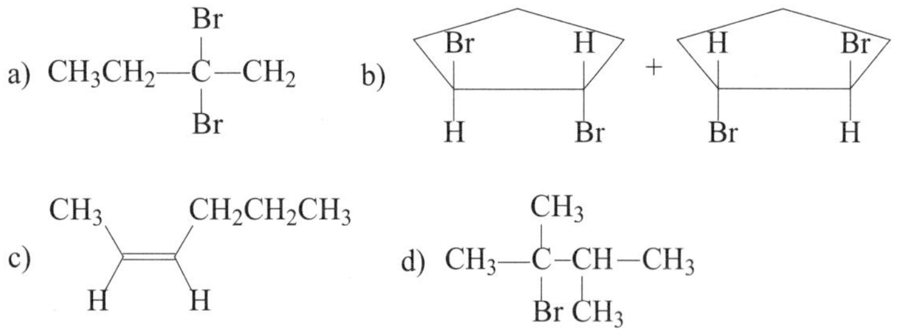

☐☐ **Hydration of an Alkene gives an Alcohol as the product, whereas hydration of an Alkyne yields either an Aldehyde or a Ketone. Why is there such a difference?**

Hydration of the Alkyne breaks only one of the pi bonds, leaving an enol as the product. Enols quickly tautomerize to form carbonyl compounds.

☐☐ **What is the Lindlar Catalyst?**

Palladium on Calcium Carbonate that has been poisoned with Lead Acetate and Quinoline.

❏❏ **What is the Lindlar Catalyst used for?**

In a catalytic hydrogenation reaction, it is used to convert Alkynes to cis Alkenes by syn addition of H_2 to the triple bond.

❏❏ **Why is a special catalyst needed for this transformation?**

Because under normal catalytic hydrogenation conditions (High pressures of H_2 with a transition metal catalyst) the Alkene resulting from reduction of the Alkyne is itself reduced to an Alkane. The deactivated Lindlar Catalyst prevents this reaction.

❏❏ **Are there any other catalysts that can be used for this reaction?**

Yes, Nickel Boride and Palladium on Barium Sulfate can also be used.

❏❏ **How do you convert Alkynes to trans Alkenes?**

By reduction of the Alkyne with Lithium, Sodium, or Potassium metal in liquid Ammonia.

❏❏ **What would be the IUPAC names of the following compounds?**

a) cyclopentyl-C≡CH

b) structure with two CH₃ groups, ethyl, connected to C≡C-CH₃

c) HC—C≡C—C≡C—CH

d) H₂C=CHCH₂CH₂—C≡CCH₂CH₃

(a) 1-Ethynylcyclopentane, (b) 6,6-Dimethyl-2-octyne, (c) 2,4-Hexadiyne, (d) 1-Octen-5-yne

❏❏ **Cyclononyne is a stable molecule, while cyclooctyne is reactive and polymerizes on standing and smaller cycloalkynes cannot even be isolated at room temperature. Why is this?**

Rings smaller than eight Carbons can not accommodate the 180 degree bond angles around the sp hybridized Carbons of an Alkyne. The Alkyne essentially requires four Carbons to lie on a straight line. The larger ring size of the cyclooctyne molecule allows the material to be isolated, but the small ring size still puts a lot of angle strain on the (ideally) 180 degree bonds of the sp hybridized Carbons, making the molecule reactive. The slightly larger ring size of cyclononyne relieves some of this stress. Larger rings would be expected to be even more stable.

❏❏ **What are the structures of the following molecules?**

a) 3-Methyl-4-vinylcyclopentene
b) (2E, 4E, 6E)-3-Chlorooctatriene
c) 3-Propyl-1-hexyne

❏❏ **1-Butyne is a gas (b. pt. = 8 °C), while 1-Pentyne is a liquid (b. pt. = 40 °C). Why is this?**

Like all hydrocarbons, Alkynes are nonpolar. Only van derWaals forces hold the molecules together. These forces increase with molecular size, so the larger pentyne has a higher boiling point than the smaller butyne.

❏❏ **How do you recognize whether an organic reaction is an oxidation, a reduction, or neither?**

Write a balanced half-reaction by using the following rules:

1) Write the half-reaction using the organic starting materials and products
2) Balance the numbers and types of atoms on each side. If the reaction is being run in acid, use H^+ to balance Hydrogens and H_2O to balance Oxygens. If it is being run in basic solution, use H_2O and OH^-.
3) Balance the charge by adding electrons to the appropriate side. If the electrons appear on the right hand side of the equation, it's an oxidation. If they appear on the left-hand side it's a reduction. If there are no electrons needed to balance charge, the reaction is neither an oxidation nor a reduction.

❏❏ **In respect to Alkenes, what is Osmium tetroxide used for?**

It is used to convert Alkenes to glycols.

❏❏ **How is this usually accomplished?**

The Alkene is reacted with one equivalent of Osmium tetroxide, followed by treatment with Sodium bisulfite in water.

❏❏ **What is the stereochemistry of this reaction.**

The reaction is syn stereoselective.

❏❏ **What is the initial product when Osmium tetroxide is reacted with an Alkene?**

The product is a cyclic osmate ester.

❏❏ **If Osmium tertroxide were reacted with cyclohexene, what would the product be?**

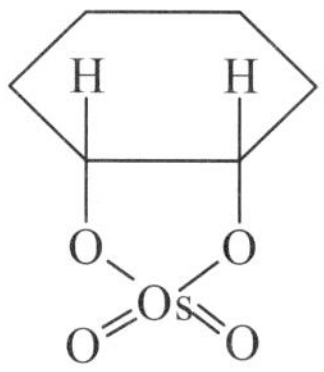

[Structure: cyclohexane ring with two H atoms and a cyclic osmate ester group (O-Os(=O)(=O)-O)]

❏❏ **What is usually done with the osmate esters that result from the reaction between Osmium tetroxide and an Alkene?**

The esters are not isolated, but rather are directly treated with a reducing agent such as $NaHSO_3$, which converts the esters to cis-glycols.

❏❏ **What are the drawbacks of reactions using Osmium Tetroxide?**

Osmium Tetroxide is both toxic and expensive.

❏❏ **What is ozonolysis?**

Treatment of an Alkene with Ozone followed by an appropriate workup. The process results in cleavage of the C-C double bond and leaves two carbonyls in its place.

❏❏ **What is the usual workup after treatment with Ozone?**

Treatment with a weak reducing agent, such as dimethyl sulfide $(CH_3)_2S$ or zinc.

❑❑ What does the ozonolysis intermediate look like?

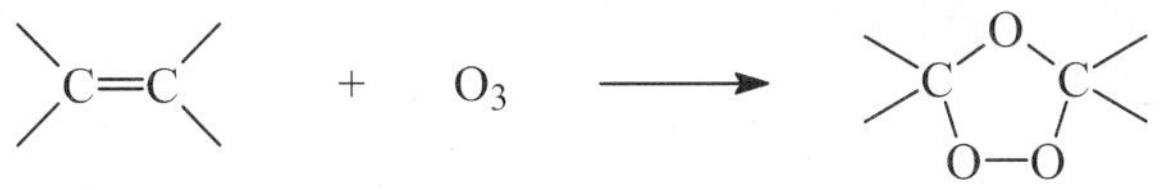

$$\text{C=C} + O_3 \longrightarrow \text{an ozonide}$$

❑❑ Ozonides are hydrolyzed to give carbonyl compounds simply by adding water, yet Zinc metal or dimethyl sulfide is often included in the hydrolysis step. Why is this?

Under the conditions of ozonolysis, any Aldehyde products would be easily oxidized to Carboxylic Acids. Zinc or dimethyl sulfide react with any oxidizing agents present and prevent this.

❑❑ Ozonolysis of an unknown Alkene yields two equivalents of the following molecule. What is the structure of the Alkene?

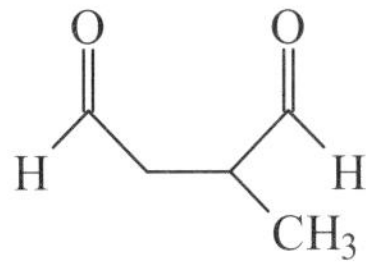

There are two possible structures for the Alkene:

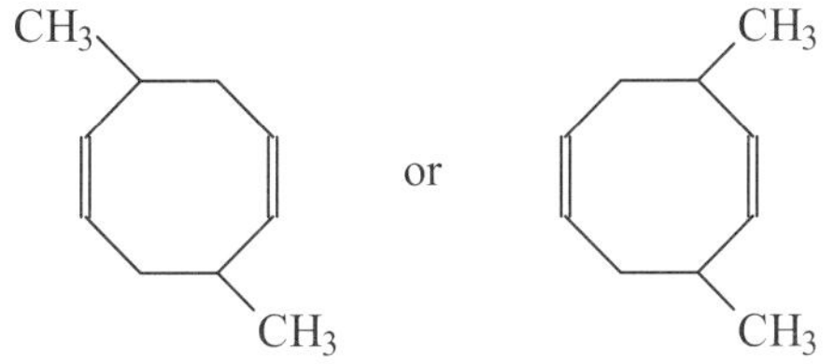

❑❑ How may Alkynes be prepared from Alkenes?

Treatment of an Alkene with Bromine or Chlorine gives a vicinal dihalide. The dihalide may be doubly dehalogenated with a strong base (such as Sodium Amide, $NaNH_2$) to give the Alkyne.

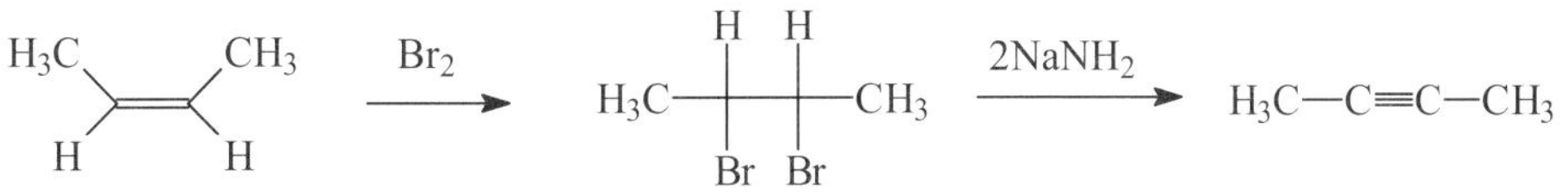

$$H_3C-CH=CH-CH_3 \xrightarrow{Br_2} H_3C-CHBr-CHBr-CH_3 \xrightarrow{2NaNH_2} H_3C-C\equiv C-CH_3$$

❑❑ When 1-Pentene is used to prepare 1-Pentyne by the above reaction, three equivalents of Sodium Amide are required rather than two. Why is this?

The Alkyne that results from 1-Pentene is a terminal Alkyne, and therefore acidic. As soon as it forms, it reacts with Sodium Amide to form an anion. Overall this consumes one equivalent of Sodium Amide, so three equivalents total are required to perform the double dehalogenation. Final isolation of the Alkyne requires treatment with water to reprotonate the Alkyne salt.

❑❑ When Propene is treated with Bromine and then doubly dehalogenated with Sodium Amide, a minor product is often observed. What could this product be and how could it form?

The minor product formed is allene. The alkenyl halide that results from the first dehydrohalogenation step can lose the next proton from two possible positions (a and b). Loss of proton (a) results in Allene, loss of proton (b) gives the expected Alkyne. The Alkyne product is preferred.

$H_2C=C=CH_2$ ← (a) — Alkenyl Halide — (b) → $H_3C-C\equiv CH$

Allene   Alkyne

❏❏ **What is the result of treating an internal Alkyne with Borane followed by Acetic acid?**

The Alkyne is reduced to an Alkene.

❏❏ **What is the steroselectivity of this reaction?**

The reaction yields cis Alkenes. As a result it provides an alternative to catalytic hydrogenation using Lindlar catalyst.

❏❏ **What intermediate is obtained after treatment of the Alkyne with Borane?**

Each equivalent of Borane can react with three equivalents of Alkyne, so the intermediate is a trialkenylborane.

❏❏ **What is the regioselectivity of the Borane addition?**

For internal Alkynes there is essentially no regioselectivity, but for terminal Alkynes the Boron adds to the least substituted Carbon. Addition is therefore anti-Markovnikov.

❏❏ **Hydroboration of 2-Hexyne followed by treatment with Hydrogen peroxide in aqueous Sodium hydroxide gives two products. What could these products be, and why would there be two of them?**

The two products are 2-Propanone and 3-Propanone. Two products are obtained because the two Carbons of the Alkyne have equivalent substitution. Thus addition of the Borane is not regioselective, and the Boron may be attached to either Carbon. When the borane intermediates are treated with H_2O_2/NaOH, the Boron atoms are replaced with hydroxyl groups. Each of the two resulting enols tautomerize into a different Ketone, resulting in the two products.

❏❏ **What is the stereochemistry and regiochemistry of the hydroboration of Alkenes?**

Addition of Borane is anti-Markovnikov and syn.

❏❏ **What is the mechanism of the addition of Borane to an Alkene (use Propene as an example)?**

The reaction occurs in one step, with the bond to the Boron forming at the same time that the Hydrogen is transferred.

❏❏ **Why is the regiochemistry of hydroboration anti-Markovnikov?**

When the boron initially adds to one of the Carbons of the double bond, a temporary positive charge is created on the other one. The boron adds to the least substituted Carbon so that the positive charge will be on the most substituted (most stabilized) Carbon. The Hydrogen ends up on the Carbon with the least Hydrogens, making the addition anti-Markovnikov.

❏❏ **What is a terpene?**

A terpene is a compound whose Carbon skeleton can be broken down into two or more units that are identical with the Carbon skeleton of isoprene.

$$H_2C=\underset{\underset{\text{Isoprene}}{}}{\overset{\overset{CH_3}{|}}{C}}-CH=CH_2$$

❏❏ **How are the isoprene units in a terpene arranged?**

The isoprene units in a terpene are always strung head to tail.

tail C—C—C—C head   **Menthol**

a terpene unit

❏❏ **What is the mechanism for addition of HCl to an Alkene? Use 1-Butene as an example.**

The reaction occurs in two steps. The proton is added to the least substituted Carbon of the double bond, resulting in a carbocation intermediate. The carbocation then reacts with the chloride ion to give the final product.

$$H_2C=CH-CH_2CH_3 \xrightarrow{H-Cl} H_3C-\overset{+}{C}H-CH_2CH_3 \quad Cl^-$$

$$H_3C-\overset{+}{C}H-CH_2CH_3 \xrightarrow{Cl^-} H_3C-\underset{\underset{}{|}}{\overset{\overset{Cl}{|}}{C}}H-CH_2CH_3$$

❏❏ **Why does the proton add to the least substituted Carbon?**

Because the other Carbon of the double bond ends up having a positive charge. This positive charge is stabilized by Carbon substitution. Essentially, the proton preferentially adds to the Carbon that gives the most stable carbocation.

❏❏ **Why does Carbon substitution effect the stability of a carbocation?**

A system carrying a charge is more stable if that charge is spread out over a larger area - if it is delocalized. Alkyl groups attached to a carbocation are electron releasing through induction and hyperconjugation, thus spreading the charge out over a larger area. The more alkyl groups attached to the carbocation, the larger the area over which the charge is spread.

❏❏ **How are carbocations classified?**

They are classified according to the number of alkyl groups attached.

```
    H           R           R           R
    |           |           |           |
  +C—H        +C—H        +C—R        +C—R
    |           |           |           |
    H           H           H           R

  Methyl     Primary    Secondary    Tertiary
```

❏❏ **What would the energy diagram look like for the reaction between HCl and 1-Butene?**

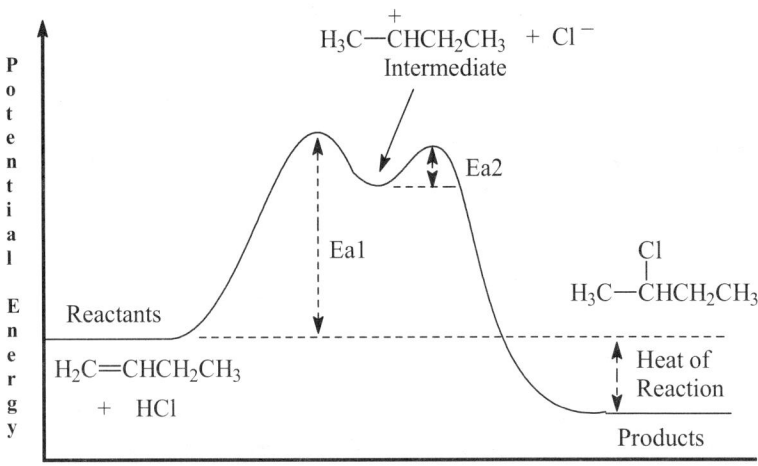

Reaction Coordinate

❏❏ **Why does the stability of the carbocation effect the regiochemistry of the product in the addition of HCl to an Alkene?**

When H^+ adds to a double bond, there are two possible products: the one in which the carbocation is on the most substituted Carbon (the more stable, lower energy carbocation), and the one in which the carbocation is on the less substituted Carbon (the less stable, higher energy carbocation). If we think of these two possibilities as competing reactions, the one in which the higher energy carbocation is formed should have the higher activation energy. Higher activation energy means a slower reaction. The more stable carbocation is formed much faster and goes on to form the observed product - the one resulting from Markovnikov addition of HCl.

❏❏ **What intermediate is formed when Bromine is reacted with an Alkene?**

A cyclic bromonium ion.

```
        Br
       / + \
      C-----C
```

❏❏ **Why does the bromonium ion form rather than a normal carbocation?**

Because in the bromonium ion the positive charge is actually spread out over three atoms (the two Alkene Carbons and the Bromine) and is therefore more stable than a localized carbocation.

❏❏ **Addition of Chlorine or Bromine to a double bond occurs with anti stereoselectivity. Why is this?**

The first step of the addition results in the formation of a bridged halonium ion. The large Halogen atom present in the halonium ion effectively blocks one face of the former Alkene. The halide ion that attacks in the second step must therefore approach from the other side, resulting in anti addition.

❏❏ **What do addition of Chlorine to an Alkene and formation of a Chlorohydrin have in common?**

Both reactions begin with the initial attack of Chlorine on the double bond to give a bridged chloronium ion.

❏❏ **Bromine in Carbon tetrachloride has been used to test for the presence of double bonds in a molecule. How does this work?**

A solution of Bromine in Carbon tetrachloride is reddish orange. When an Alkene is added to the solution, the Bromine reacts with the double bond to give a colorless dibromide compound. The loss of color indicates the presence of a double bond.

❏❏ **What is the stereoselectivity when Bromine reacts with an Alkyne?**

The stereoselectivity is the same as when Bromine adds to an Alkene - anti. This can only be observed, however, if the reaction is stopped after the addition of one equivalent of Bromine.

❏❏ **What intermediate is formed when Bromine adds to an Alkyne?**

The intermediate is similar to the one formed when Bromine reacts with Alkenes. It is a cyclic bromonium ion, but with a double bond connecting the two Carbons.

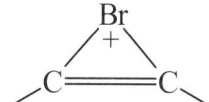

❏❏ **The addition of Halogens to Alkynes is much slower than the addition of Halogens to Alkenes. Why is this?**

The bridged intermediate formed in the reaction with Alkynes is much less stable than the one formed in the Alkene reactions due to the strain caused by the double bond in the three membered rings. The higher energy intermediate results in higher activation energy for the Alkyne reaction, and therefore a slower rate.

❏❏ **What is the mechanism for the formation of a Halohydrin?**

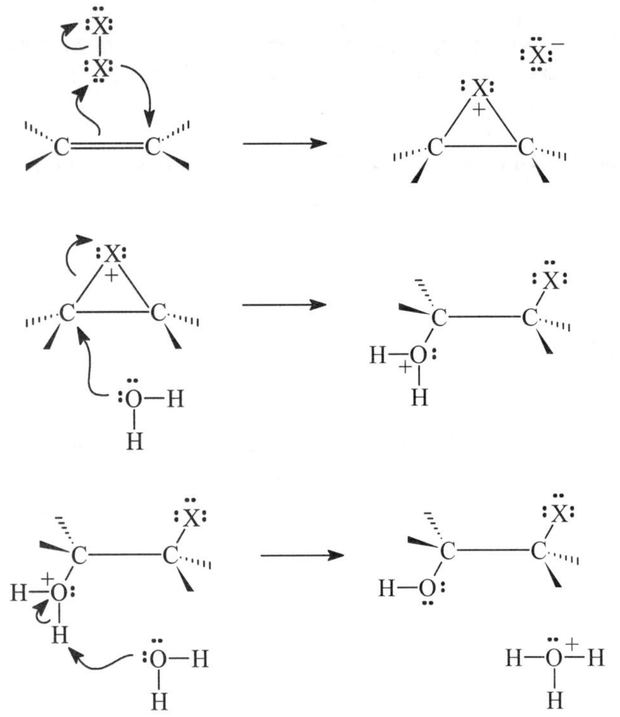

❏❏ **Ozonolysis of Alkenes produces Aldehydes and Ketones. What products result from the ozonolysis of Alkynes?**

Carboxylic acids.

❏❏ **The isomeric Alkenes 2,3-Dimethyl-1-butene, 2-Methyl-2-pentene, and 2,3-Dimethyl-2-butene have the heats of hydrogenation shown below. What conclusion might we draw from this data?**

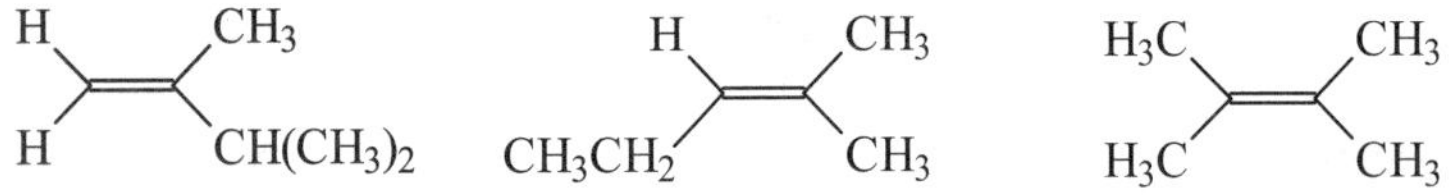

27.8 kcal/mol	26.7 kcal/mol	26.4 kcal/mol
116 kJ/mol	112 kJ/mol	110 kJ/mol

Since the heats of hydrogenation decrease with increasing substitution, we might conclude that tetrasubstituted double bonds are more stable than disubstituted double bonds, which are more stable than monosubstituted double bonds. Essentially, double bond stability increases with increasing substitution.

☐☐ **What would be the primary organic product expected in each of the following reactions?**

a) H—C≡C—CH$_2$CH$_3$ + HCl (excess) ⟶

b) 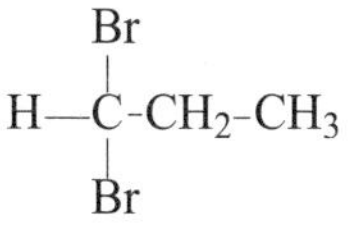 H—C(Br)(Br)-CH$_2$-CH$_3$ + NaNH$_2$ (excess) ⟶

c) 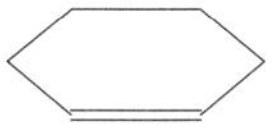 cyclohexene + Br$_2$ $\xrightarrow{H_2O}$

a) 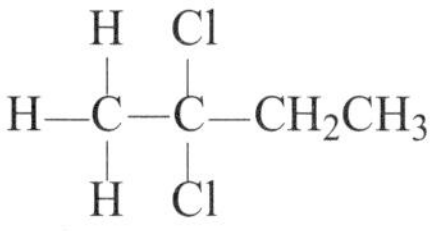 H—C(H)(H)—C(Cl)(Cl)—CH$_2$CH$_3$ b) H—C≡C—CH$_3$ c) 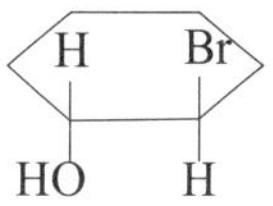 cyclohexane with H, Br, HO, H substituents

☐☐ **Why is Acetylene such a useful starting material for the production of other Alkynes?**

Acetylene can be deprotonated with Sodium Amide to give an acetylide ion. Acetylide ions can be alkylated with methyl and primary halides to give more complicated terminal Alkynes. These terminal Alkynes can in turn be deprotonated and alkylated to give internal Alkynes.

H—C≡C—H $\xrightarrow{NaNH_2}$ H—C≡C$^-$ $\xrightarrow[\text{R = methyl or 1°}]{RX}$ H—C≡C—R

H—C≡C—R $\xrightarrow{NaNH_2}$ $^-$C≡C—R $\xrightarrow[\text{R = methyl or 1°}]{R'X}$ R'—C≡C—R

☐☐ **The rate of reaction of the Hydrogen Halides with Alkenes goes in the order HF << HCl < HBr < HI. Why is this?**

The first step in the addition of a Hydrogen Halide with an Alkene is the transfer of a proton. It would be expected that the more acidic the Hydrogen Halide, the faster it would react. The order above reflects the acidity of the Hydrogen Halides: HI is the strongest acid, HBr is a bit weaker, HCl is weaker still, and HF the weakest acid.

☐☐ **When HBr is added to an Alkene, sometimes the addition is Markovnikov, and sometimes it is anti-Markovnikov. What is the difference between these two cases?**

When the addition occurs in an anti-Markovnikov fashion this indicates that peroxides are present.

❑❑ **What is a conjugated diene?**

A conjugated diene is one in which the two double bonds are joined by a C-C single bond. This structure allows interaction between the π systems of the four sp^2 hybridized Carbons.

C=C—C=C

❑❑ **What is a cumulated diene?**

A cumulated diene is one in which the two double bonds are joined through a single Carbon.

C=C=C

❑❑ **What is the structure of Allene?**

Allene is the simplest cumulated diene and has the following structure:

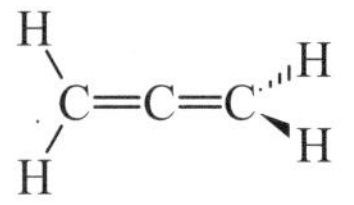

$$\begin{array}{c} H \\ \diagdown \\ C=C=C \\ \diagup \diagdown \\ H H \end{array} \begin{array}{c} H \\ \diagup \\ \\ \\ \end{array}$$

❑❑ **What is the hybridization of the central Carbon of allene?**

The central Carbon of Allene is sp hybridized.

❑❑ **What is the most stable arrangement of double bonds in a diene, isolated, conjugated, or cumulated?**

Conjugated double bonds are the most stable, followed by isolated double bonds. Conjugation usually increases the stability of a molecule by about 3.5 - 4.0 kcal/mol. Cumulated double bonds are relatively high in energy, and therefore the least stable.

❑❑ **What factor is responsible for the increased stability of conjugated double bonds?**

The factor most responsible for the higher stability of conjugated double bonds is the increased delocalization of the π electrons.

❑❑ **Draw the s-Cis and s-Trans conformations of 1,3-Butadiene.**

s-Trans form of 1,3-Butadiene

s-Cis form of 1,3-Butadiene

❑❑ **Which of these conformations is more stable?**

The s-trans conformation is the most stable. The s-cis form has an unfavorable steric interaction between Hydrogens on the two end Carbons.

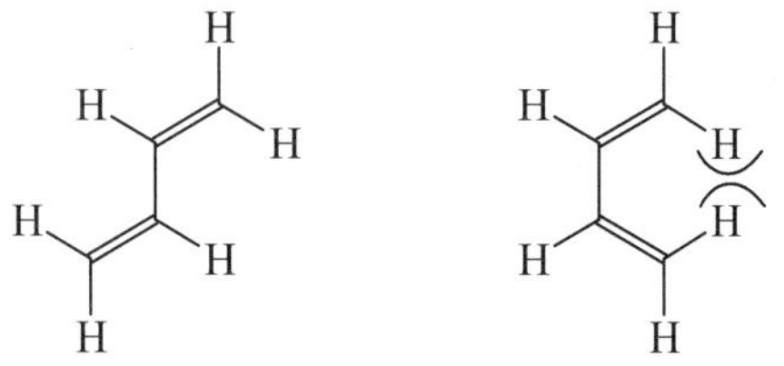

More stable   Less stable

☐☐ **Interconversion of the s-cis and s-trans confomations of 1,3-Butadiene has a large barrier (about 3.9 kcal/mol.) Why is this?**

Both the s-cis and s-trans conformations of 1,3-Butadiene are planar, with good overlap between the π orbitals of the two double bonds. This overlap results in enhanced delocalization of the π electrons due to conjugation, which lowers the energies of these two conformations. Interconversion between the two, however, requires that rotation around the central σ bond occur, disrupting the conjugation and raising the energy of the molecule. The maximum occurs when the orbitals of the two π systems are perpendicular to one another.

Lowest Energy	Highest Energy	Low Energy
Conjugated	Conjugation Destroyed	Conjugated

☐☐ **When one equivalent of HBr is reacted with 1,3-Butadiene in the presence of a radical inhibitor, the major product is 3-Bromo-1-butene. But a significant amount of 1-Bromo-2-butene is also obtained. How is this possible?**

Initial addition of a proton to one of the double bonds of 1,3-Butadiene gives an allylic carbocation. If the bromide ion reacts with this carbocation, the result is known as 1,2 addition. This gives the major product, 3-Bromo-1-butene. However, the initially formed allylic carbocation is in resonance with another form, in which the carbocation resides on the other terminal Carbon. If the bromide attacks at this center, the result is 1, 4 addition, which gives the minor product, 1-Bromo-2-butene.

$H_2C{=}CHCH{=}CH_2 \;+\; HBr \;\longrightarrow\; H_2C{=}CHCH\overset{H}{\underset{|}{-}}\overset{+}{C}H_2$

$H_2C{=}CH\overset{+}{C}H{-}\overset{H}{\underset{|}{C}}H_2 \;+\; Br^- \;\longrightarrow\; H_2C{=}CH\overset{Br}{\underset{|}{C}}H{-}\overset{H}{\underset{|}{C}}H_2$ 1,2 Addition Major product

$\left[\; H_2C{=}CH\overset{+}{C}H{-}\overset{H}{\underset{|}{C}}H_2 \;\longleftrightarrow\; \overset{+}{H_2C}{-}CH{=}CH{-}\overset{H}{\underset{|}{C}}H_2 \;\right]$ Two resonance forms, both allylic carbocations

$\overset{+}{H_2C}{-}CH{=}CH{-}\overset{H}{\underset{|}{C}}H_2 \;+\; Br^- \;\longrightarrow\; \overset{Br}{\underset{|}{H_2C}}{-}CH{=}CH{-}\overset{H}{\underset{|}{C}}H_2$ 1,4 Addition Minor product

☐☐ **What is a Diels-Alder reaction?**

It is a cylcoaddition reaction in which conjugated dienes react with certain types of Carbon-Carbon double or triple bond compounds to form a cyclohexane derivative.

The arrows are a formalism. The reaction is believed to occur in an entirely concerted fashion.

☐☐ **What is the major requirement for a diene to be able to undergo a Diels-Alder reaction?**

It must be able to assume an s-cis conformation.

☐☐ **By what name is the product of a Diels-Alder reaction known?**

These products are often referred to as Diels-Alder adducts.

☐☐ **By what name is the double or triple bonded compound that reacts with the Diene in a Diels-Alder reaction known?**

It is called the Dienophile.

☐☐ **The simplest Diels-Alder reaction, between 1,3-Butadiene and Ethylene, is very slow and requires high temperatures and pressures to occur. What changes can be made to this reaction to increase the rate?**

The rate of Diels-Alder reactions increase if electron-withdrawing substituents are added to one reactant and electron donating substituents are added to the other. Addition of electron donating or withdrawing substituents to just one of the reactants will increase the rate, but not by as much.

☐☐ **Diels-Alder reactions can be used to construct complicated bicyclic systems. What products are possible when two molecules of 1,3-Cyclohexadiene react?**

There are two possible isomers that can result from the this Diels-Alder reaction: the exo isomer, in which dienophile ring is on the same side as the bridge Carbons, or the endo isomer, in which the dienophile ring is on the side opposite the bridge Carbons. The endo adduct is the preferred one under normal conditions.

endo isomer   exo isomer
                     (usually the minor isomer)

☐☐ **How is the configuration of the Dienophile effected by a Diels-Alder reaction?**

The configuration of the Dienophile is retained in the product: any substituents on the Dienophile that were cis to one another remain cis in the adduct, any groups that were trans remain trans.

☐☐ **What is a Pericyclic reaction?**

A Pericyclic reaction is a reaction that involves a cyclic redistribution of bonding electrons. The reaction occurs in one concerted step, with no intermediates.

STEREOCHEMISTRY PEARLS

The difference between a genius and a lunatic is that the genius has proof.
Dominique Bouchard

❏❏ **What is meant by the term Stereochemistry?**

Stereochemistry is the study of the relationship between a molecule's physical and chemical properties and the arrangement of its atoms in space. It's the study of the three dimensional structures of molecules.

❏❏ **What are Stereoisomers?**

Stereoisomers are isomers that have the same constitution, but which differ in the geometric orientation of their atoms in space.

❏❏ **Who first proposed the idea of Stereoisomers?**

Jacobus van't Hoff and Charles Le Bel in 1874 independently proposed that the four bonds of Carbon were directed towards the corners of a tetrahedron, and that therefore two compounds might be different because of the arrangement of their atoms in space.

❏❏ **What is optical activity?**

Optical activity is the ability of a substance to rotate the plane of polarized light as the light passes through it.

❏❏ **What is plane polarized light?**

When normal light travels through space, the waves of the electric field vibrate in all of the planes perpendicular to the direction of travel. Polarized light is light that vibrates in only a single plane. All of the waves vibrating in other planes have been filtered out.

❏❏ **What is a Chiral Center?**

A Chiral Center is a tetrahedral atom with four different unique substituents attached.

❏❏ **Is the presence of a Chiral Center necessary for a molecule to be chiral?**

No. A Chiral Center is neither necessary nor sufficient for a molecule to be chiral.

❏❏ **Is Carbon the only atom that can form a Chiral Center?**

No. Any atom capable of forming four (or more) bonds can potentially be a Chiral Center. A Quaternary Nitrogen, for example, could be chiral with the proper substitution.

❏❏ **What are the various terms that have been used to describe a chiral Carbon?**

Chiral Center, assymmetrc center, stereocenter, and stereogenic center are all common.

❏❏ **How is optical activity measured?**

Optical activity is measured using an instrument called a polarimeter.

❏❏ **What would a general diagram of a polarimeter look like?**

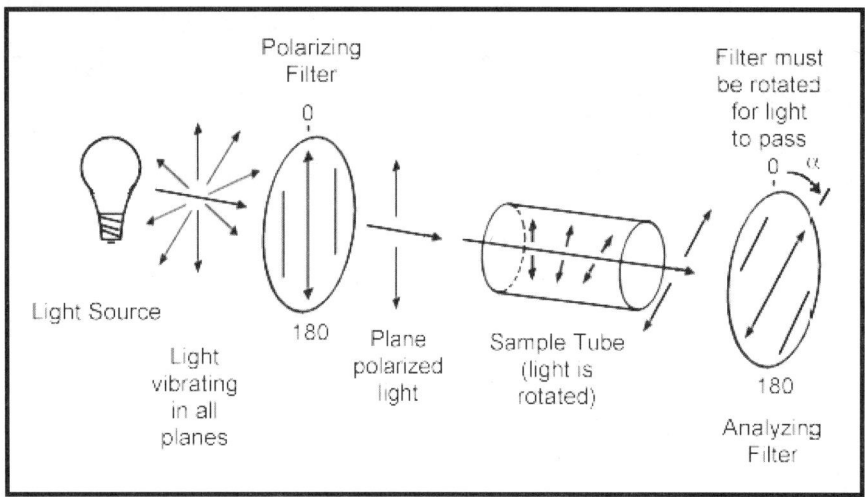

❏❏ **What do dextrorotatory and levorotatory mean?**

Dextrorotatory is the term applied to compounds that rotate polarized light in a clockwise fashion, and levorotatory refers to compounds that rotate polarized light in a counter-clockwise fashion.

❏❏ **What is α?**

The observed rotation, α, is the number of degrees through which the analyzing filter of a polarimeter must be rotated

❏❏ **Aside from the structure of the molecule, what factors can effect the observed optical rotation of a pure compound?**

The concentration of the sample and the length of the sample tube would have the greatest effect. To a lesser degree, the temperature of the sample and the wavelength of the light used for the measurement may also have an effect.

❏❏ **What is the Specific rotation of a compound?**

The Specific rotation of a compound, [α], is defined as the observed rotation of a compound at a specific cell (sample tube) length and at a specific sample concentration. The standard cell length is 1 decimeter (10 cm). The standard concentration of a sample is 1 g/mL. Since these conditions are not always possible, specific rotation can be calculated as:

$$\text{Specific rotation} = [\alpha]_\lambda^T = \frac{\text{observed rotation (degrees)}}{\text{length (dm) x concentration (g/mL)}}$$

For liquid samples, the concentration is equal to the density of the liquid. The temperature in degrees celsius (T), and the wavelength of light used for the measurement (λ) are designated as a superscript and subscript, respectively.

❏❏ **What is meant by** $[\alpha]_D^{25}$ **?**

This is the specific rotation of a molecule measured at 25 degrees celsius, using polarized light of the wavelength corresponding to the D line of sodium (λ = 589 nm).

❏❏ **A pure sample of an optically active compound is dissolved in solvent at a concentration of 0.32 g/mL. If the observed rotation using a 1 dm sample tube is +3.23 degrees, what is the specific rotation of the compound?**

[α] = +10.1 degrees.

❏❏ **What is required for a molecule to possess optical activity?**

It must be chiral, meaning that it must be non-superposable on its mirror image.

❏❏ **What are Enantiomers?**

Enantiomers are stereisomers that are related as non-superposable mirror images. A pair of gloves is a good example of Enantiomers.

❏❏ **Are there any other types of Stereoisomers?**

Yes, Diastereomers, which are Stereoisomers that are not mirror images of one another.

❏❏ **Does a compound have to have a Chiral Center to have stereochemistry and Stereoisomers?**

Absolutely not. Alkenes such as 2-Butene have cis/trans stereochemistry. *Cis*-2-Butene and *trans*-2-Butene are Diastereomers. Other example include 1,4-disubstituted Cycloalkanes, which can also exhibit cis/trans isomerism.

❏❏ **How do the properties of a pair of Enantiomers compare?**

All of the physical and chemical properties (melting point, boiling point, solubility, reactivity, etc.) of Enantiomers are identical. They differ only in the direction in which they rotate plane polarized light and in how they interact with other pure chiral molecules.

❏❏ **How do Enantiomers differ in how they rotate polarized light?**

Enantiomers have equal and opposite specific rotations. One isomer is always levorotatory while the other is always dextrorotatory. For example, the two Enantiomers of 2-Butanol have specific rotations of +13.52 degrees and -13.52 degrees, respectively.

❏❏ **How are Enantiomers named so that they can be told apart?**

The configuration (spatial arrangement of the atoms) of a chiral molecule is designated using the R, S convention. With a pair of Enantiomers having one Chiral Center, one of the Enantiomers would have the designation R, and the other would have the designation S. For molecules having more than one Chiral Center, the designation of each center must be given.

❏❏ **How is the R/S designation of a Chiral Center assigned?**

Each of the four groups attached to a Chiral Center is assigned a priority (1-4) according to the Cahn-Ingold-Prelog system. The group of lowest priority (4) is directed away from you. The three groups of higher priority should project toward you. The three groups are then read in order from highest priority to lowest. If the reading proceeds in a clockwise direction then the designation is R, if it proceeds in a counterclockwise direction, then the designation is S.

❏❏ **How is priority assigned?**

Priority is assigned based on the atomic number of each of the atoms bonded to the Chiral Center (higher number = higher priority). If two or more of the atoms are identical, look at the next set of atoms bonded to those and continue until priority can be assigned. If an atom participates in multiple bonds it is considered to be bonded to an equivalent number of similar atoms by single bonds (ie Carbon double bonded to Oxygen is considered to be bonded to two Oxygens). If two atoms of equivalent atomic number (such as Hydrogen and Deuterium) are bonded to no other atoms, atomic mass decides priority.

❑❑ **What is the best way to go about determining the configuration of a Chiral Center when presented with a two dimensional drawing?**

Build a model that matches the structure and manipulate it in space so that the lowest priority group is directed away from you. Proceed with the assignment from there. In the absence of a model, imagining a steering wheel with three spokes can often provide assistance. Your hand can also serve well as a model when necessary, your thumb, index finger and middle finger playing the part of the three high priority groups and your forearm the part of the lowest priority group.

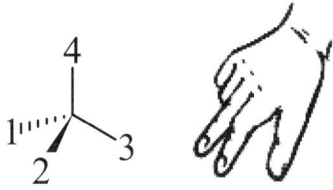

❑❑ **What is a Racemic mixture?**

A Racemic mixture is a 1:1 mixture of two Enantiomers. They have equal and opposite specific rotations, so their optical activities cancel out and the resulting mixture is optically inactive.

❑❑ **How difficult is it to separate a Racemic mixture into two pure Enantiomers?**

Since all of the properties of the two Enantiomers are identical, it can be very difficult. Normal separation approaches such as distillation and crystallization are ineffective.

❑❑ **Are all Stereoisomers difficult to separate?**

Not as difficult as a Racemic mixture. A mixture of Diastereomers is much easier to separate because Diastereomers have different chemical and physical properties (though they may be similar).

❑❑ **What is the process of separating as Racemic mixture into its component Enantiomers called?**

Resolution.

❑❑ **What is the easiest way to resolve a Racemic mixture?**

React it with another chiral compound, separate the resulting Diastereomers, and somehow reverse the reaction to regenerate the pure optically active compounds.

❑❑ **Who performed the first successful resolution of a Racemic mixture?**

Louis Pasteur, in 1848 while working with Sodium Ammonium tartrate crystals.

❑❑ **A pure sample of (S)-(-)-Malic acid has a specific rotation of -27 degrees. What would be the specific rotation of a mixture containing 68% (+) Malic acid and 32% (-) Malic acid?**

The mixture would essentially consist of 64% (2 x 32%) optically inactive racemate and 36% (+) Malic acid. (+) Malic acid should have an equal and opposite specific rotation than (-) Malic acid, so the observed specific rotation should be 36% of 27 degrees, or 9.72 degrees.

❑❑ **Which of the following objects would have a plane of symmetry?**

a) A book
b) A glass beaker
c) A ping-pong ball
d) An automobile
e) A lightbulb

a) No. The book wouldn't have a plane of symmetry because of the writing inside and because of the difference between the front and back covers. b) Maybe. The glass beaker might have a plane of symmetry if there was no writing on it. c) Yes. A ping pong ball would have a plane of symmetry along the seam between the two halves. d) No. The automobile's steering wheel, among other things, prevents it from having a plane of symmetry. e) No. The threads on the lightbulb would prevent this.

❑❑ **What changes are necessary to convert a model of a molecule into it's Enantiomer?**

Swapping any two groups on the models' Chiral Center(s) will convert it to the Enantiomer.

❑❑ **What would be the order of priority of the following groups?**

a) —C(=O)OH   b) —CH₂OH   c) —Br

d) —CH₂CH₃   e) —C(=O)H   f) —CH=CH₂

From highest to lowest priority: c, a, e, b, f, d.

❑❑ **What is the relationship between the R/S configuration of a molecule and its +/- optical rotation?**

There is no relationship whatsoever between the configuration and the optical rotation. R/S configuration is a way of indicating the three dimensional arrangement of the atoms around the Chiral Center, and is determined by a series of rules. The +/- optical rotation is an experimentally determined physical property.

❑❑ **What is a meso compound?**

A meso compound is a compound having two or more Chiral Centers, but which is optically inactive due to symmetry.

❑❑ **How do you recognize a meso compound?**

When a meso compound is arranged in its most symmetrical conformation, one half of the molecule is a mirror image of the other. The molecule has a plane of symmetry.

❑❑ **Why are meso compounds optically inactive?**

Because one half of a meso compound is a mirror image of the other, the optical activity conferred by the Chiral Centers on one side of the molecule is exactly cancelled out by the optical activity of the Chiral Centers on the other side. A meso compound is, in effect, very much like a Racemic mixture.

❏❏ **What would be the configuration of each of the following molecules?**

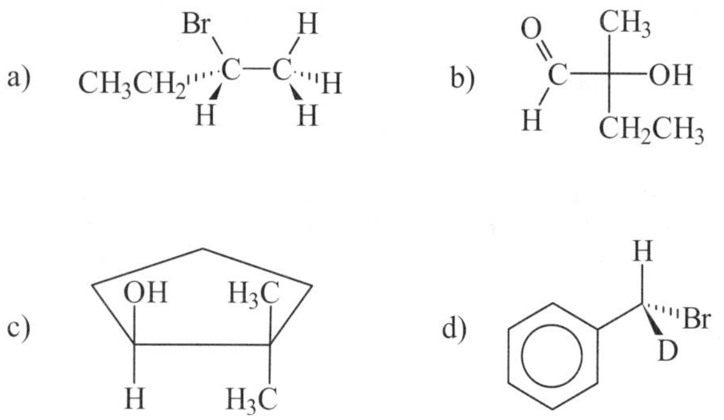

a) [structure: CH₃CH₂ and Br on one carbon, H and H on adjacent carbon with H wedges]

b) [structure: CHO, CH₃, OH, CH₂CH₃ around central C]

c) [cyclopentane with OH, H₃C, H, H₃C substituents]

d) [benzene-CH with H, Br, D]

(a) S, (b) S, (c) S, (d) R

❏❏ **Bromination of *cis*-2-Butene generates a molecule with two Chiral Centers. Would you expect the product of this reaction to be optically active?**

No, the product would not be optically active because a Racemic mixture is generated. This demonstrates a general principle: optically active products cannot be generated unless you begin with optically active starting materials or somehow have an optically active environment (such as chiral catalysts or solvents).

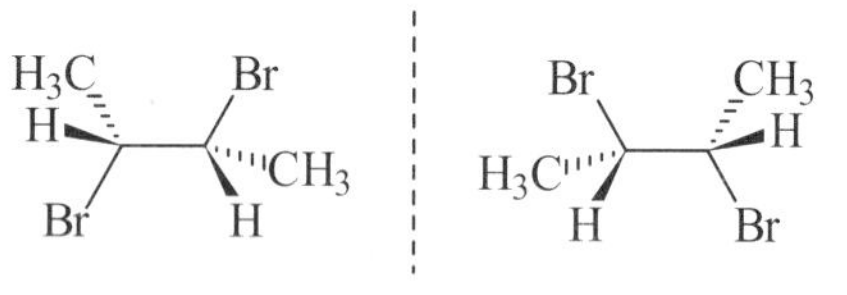

A pair of enantiomers

❏❏ **Bromination of *trans*-2-Butene gives a single molecule, rather than a mixture of Enantiomers, yet the product is still not optically active. Why is this?**

The product is a meso compound.

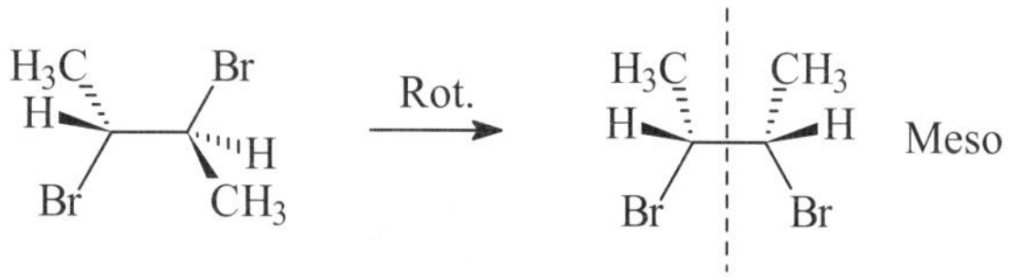

❏❏ **What is necessary for a molecule to be considered meso?**

To be considered meso, a compound must contain Chiral Centers, but be superposable on its mirror image. This is usually recognized by the presence of a plane of symmetry in the molecule.

❏❏ **What is the stereochemical relationship between each of the following pairs of molecules?**

a) 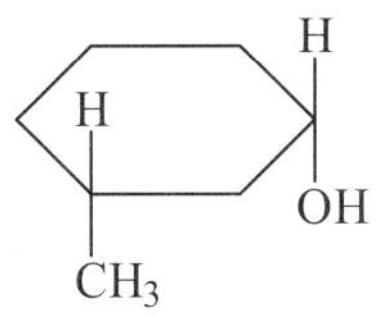 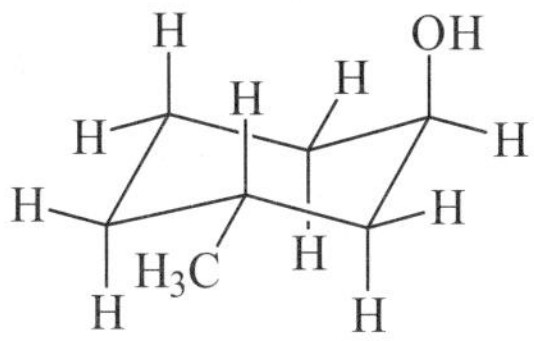 [structures of methylcyclohexanol pair]

b) 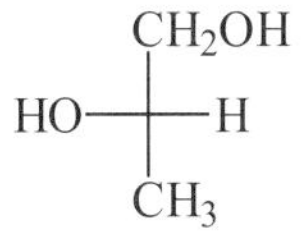 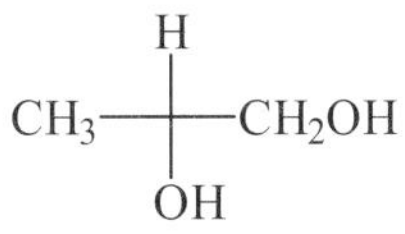 [Fischer projection pair with CH₂OH, HO, H, CH₃ and CH₃, H, OH, CH₂OH]

c) 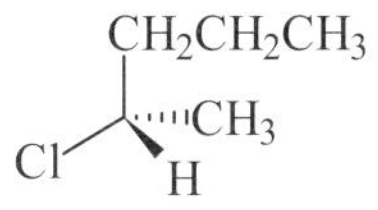 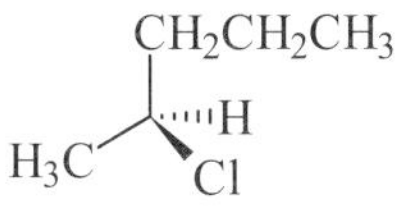 [wedge-dash structures with CH₂CH₂CH₃, Cl, CH₃, H groups]

(a) Diastereomers, (b) Identical, (c) Identical

❏❏ **Draw and name all of the Stereoisomers of 2,3,-Hexanediol.**

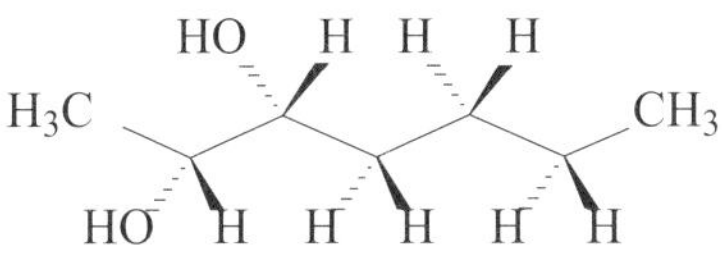

(2R, 3R)-2,3-Hexanediol

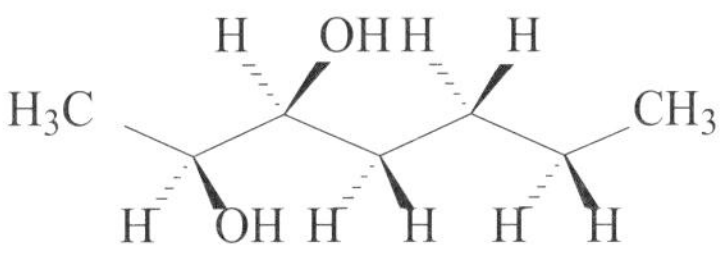

(2S, 3S)-2,3-Hexanediol

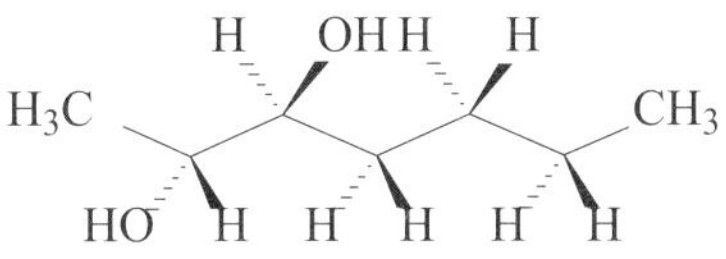

(2R, 3S)-2,3-Hexanediol

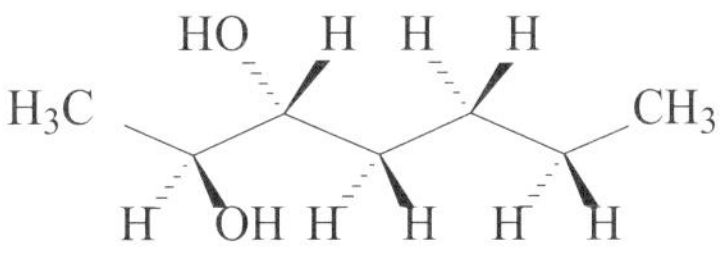

(2S, 3R)-2,3-Hexanediol

❏❏ **What is the maximum possible number of Stereoisomers of a compound containing Chiral Centers?**

For a molecule with n stereocenters, the maximum possible number of Stereoisomers is 2^n.

❏❏ **Is the maximum possible number of Stereoisomers the number of steroisomers that is always found?**

No, if meso compounds are possible the number of Stereoisomers will be fewer.

❏❏ **How do you recognize from a structure whether or not some of the Stereoisomers will be meso?**

The structure will have an even number of Chiral Centers and it will be possible to divide them up into pairs of Chiral Centers having identical substituents.

❏❏ **Which of the following molecules contain a Chiral Center?**

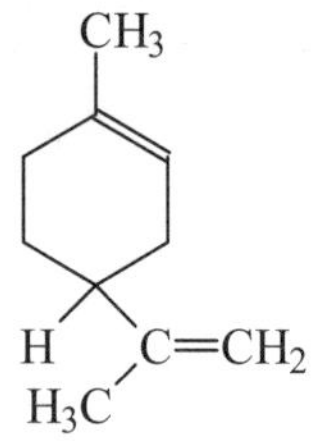

Limonene

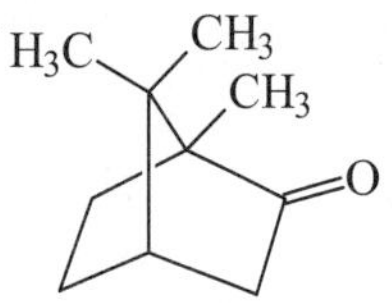

Camphor

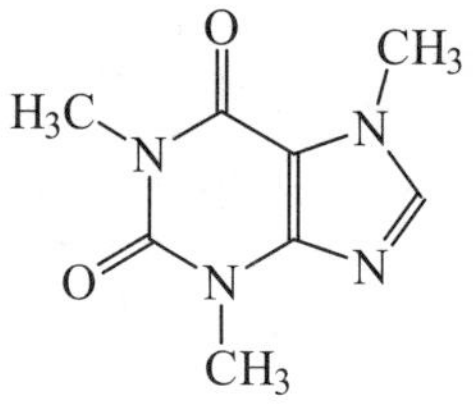

Caffeine

Limonene and Camphor both contain Chiral Centers, one for Limonene and two for Camphor. Caffeine has no Chiral Center.

❏❏ **With the exception of their rotation of plane polarized light, the two Enantiomers of Carvone have identical physical properties (melting point, boiling point, etc.). Their odors, however, are completely different. (R)-(-)-Carvone is the odor of spearmint, while (S)-(+)-Carvone is the odor of caraway seeds. How is this possible?**

The physical and chemical properties of Enantiomers are identical except when they interact with other chiral, optically active materials. The scent receptors in the human nose are chiral.

❏❏ **How many Stereoisomers are possible for each of the following molecules?**

a) 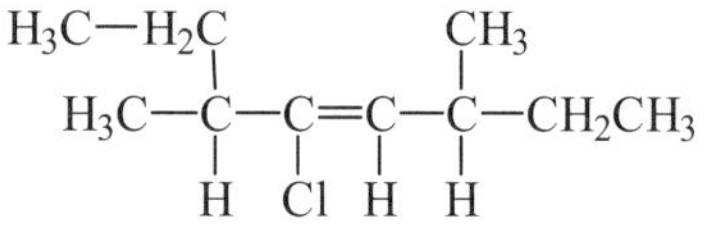   b) 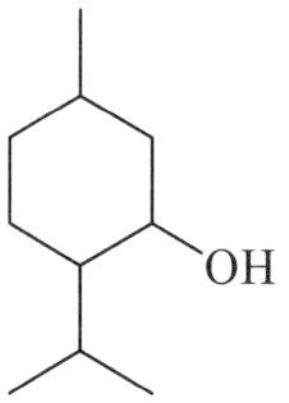   c)

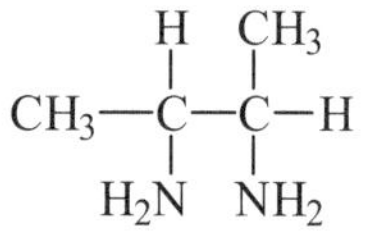

(a) 8 (two Chiral Centers and one double bond, so $n = 3$), (b) 8 (three Chiral Centers), (c) 3 (there are two Chiral Centers, but one of the isomers is a meso compound).

❏❏ **What is the maximum number of Stereoisomers possible for each of the following molecules?**

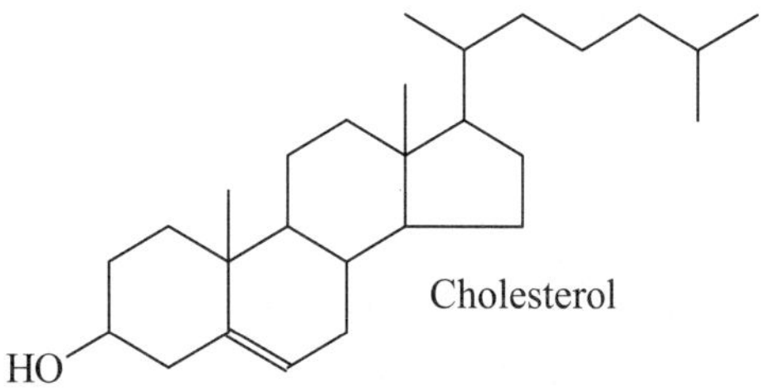

Cholesterol

Nicotine

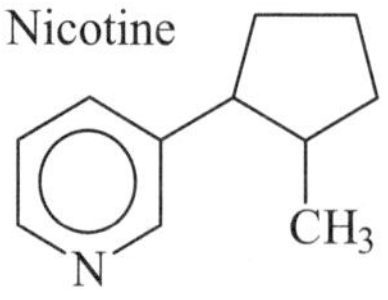

Cholesterol has 8 Chiral Centers and therefore 256 maximum possible Stereoisomers. The double bond has only one possible configuration, so it is not considered in the calculation. Nicotine has 2 Chiral Centers and therefore 4 maximum possible stereisomers.

❏❏ **By what other name is a Racemic mixture known?**

A Racemic mixture is also known as a racemic modification.

❏❏ **What would the structures look like for the following molecules?**

a) (S)-2-Methylbicyclo[2.2.1]heptane
b) (3S,4R)-3,4-Dichloro-4-methylcyclohexene
c) (2S, 4R)-4-Methyl-2-heptanol

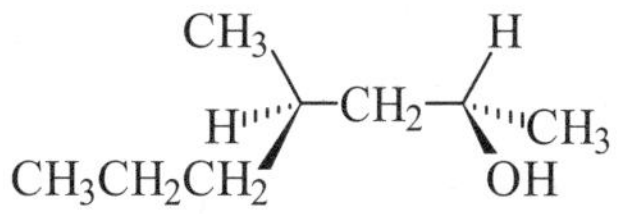

❏❏ **Which of the following compounds are meso?**

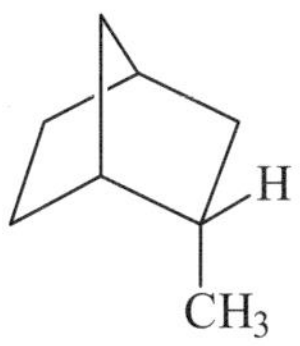

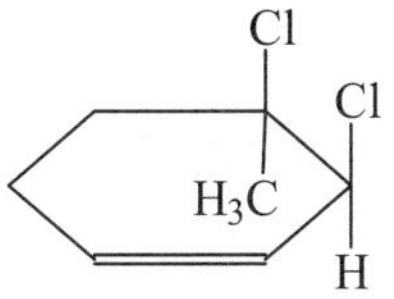

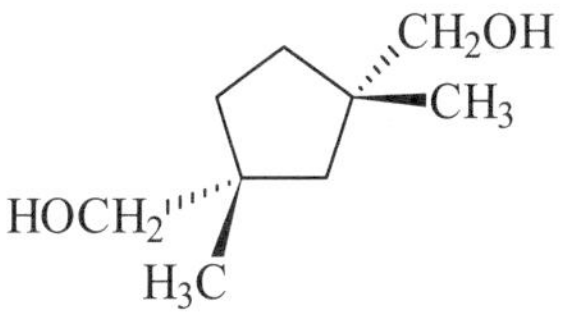

Compounds a, b, and d are meso.

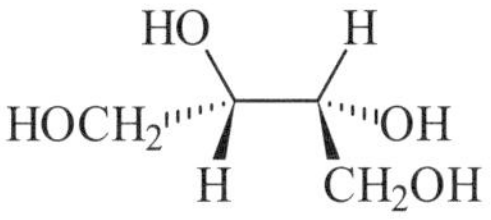

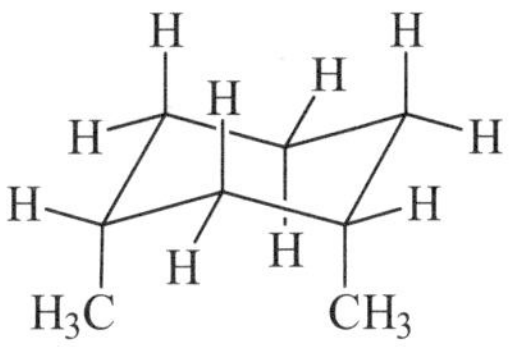

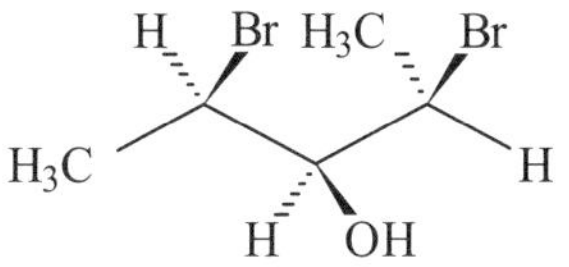

❏❏ **What is the difference between a Stereoselective reaction and a Stereospecific reaction?**

A Stereoselective reaction is one in which one Stereoisomer is formed (or destroyed) in greater amounts than all of the other possible stereoisomeric products. For example, catalytic hydrogenation of 1,2-Dimethylcyclohexene gives 85% *cis*-1,2-Dimethylcyclohexane and 15% *trans*-1,2-Dimethylcyclohexane because the Hydrogen usually adds syn. A Stereospecific reaction is one in which different Stereoisomers of the reactant give stereoisomerically different products. For example, addition of Bromine to *cis*-2-Butene gives a Racemic mixture of (2R,3R)-2,3-Dibromobutane and (2S, 3S)-2,3-Dibromobutane, whereas addition of Bromine to *trans*-2-Butene results in meso (2R, 3S)-2,3-Dibromobutane.

❏❏ **What is a Regioselective reaction?**

A Regioselective reaction is a reaction which could produce two (or more) constitutional isomers, but which gives one of them in greater amounts than the other. Addition of HCl to propene is a Regioselective reaction: addition in accordance with Markovnikov's law is preferred and results primarily in 2-Chloropropane.

❏❏ **A pure sample of (R)-(-)-Carvone (spearmint oil) has a specific rotation of -62.5 degrees. A sample of spearmint oil contaminated with its Enantiomer was found to have a specific rotation of -46.0 degrees. What is the percent optical purity of this sample?**

$$\% \text{ Optical Purity} = \frac{[\alpha] \text{ sample}}{[\alpha] \text{ pure enantiomer}} \times 100 = \frac{-46.0}{-62.5} \times 100 = 73.6 \% \text{ optical purity}$$

❑❑ **What is a Fischer projection?**

Fischer projections are a method for representing stereochemical information in a two dimensional format. The four bonds to a Chiral Center are represented by a cross. The horizontal lines represent bonds projecting toward you, and the vertical lines represent bonds projecting away. (A good way to remember this convention is the think of the handlebars of a bicycle.)

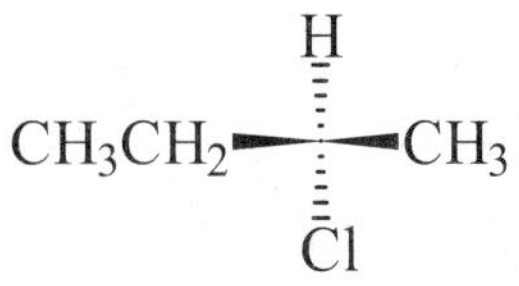

$$CH_3CH_2 \mathrel{\mathop{\rule{0pt}{0.7em}}\limits^{H}_{Cl}} CH_3 \quad \text{Becomes} \quad CH_3CH_2 \mathrel{\mathop{\rule{0pt}{0.7em}}\limits^{H}_{Cl}} CH_3$$

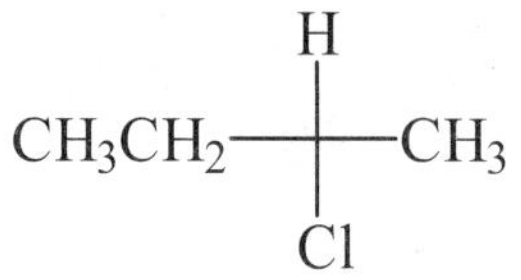

❑❑ **What are the permissible ways of manipulating Fischer projections on a page?**

Fischer projections on a page may be rotated by 180 degrees with no loss of stereochemical information. Single rotation by 90 degrees or mentally removing the projection from the page (flipping) is not allowed, however, any even numbered combination of 90 degree rotations and or flips will retain the correct configuration (ie two flips or a flip and a 90 degree rotation).

❑❑ **What happens if a Fischer projection is rotated by 90 degrees or flipped out of the plane of the page?**

Rotation by 90 degrees or flipping converts the molecule into a representation of its Enantiomer.

❑❑ **Who developed Fischer projections?**

The German chemist Emil Fischer introduced the representations in 1891.

❑❑ **What else is Emil Fischer known for?**

He won the 1902 Nobel Prize for determining the configurations of the four Chiral Carbons in the sugar Glucose.

❑❑ **How could the following molecules be represented using Fischer projections?**

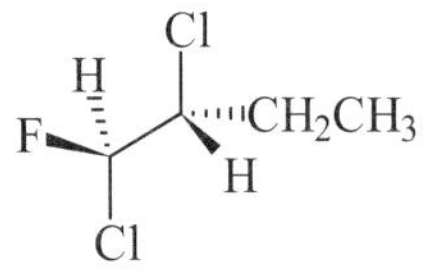

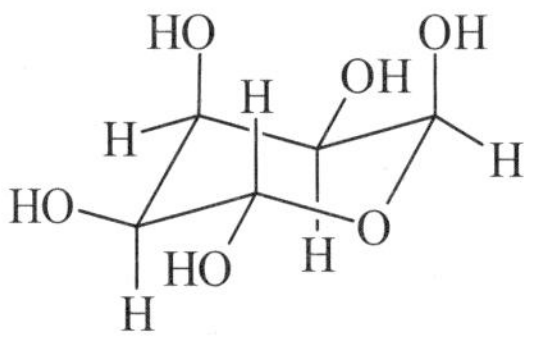

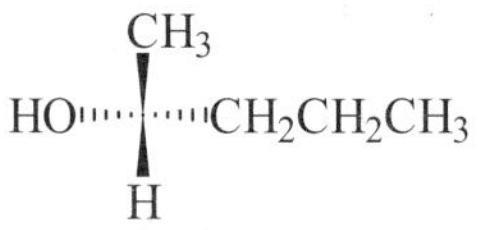

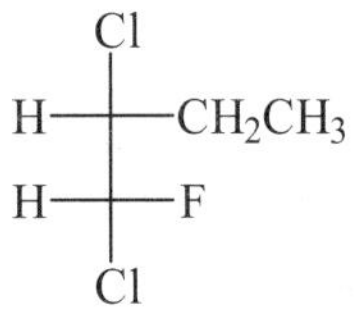

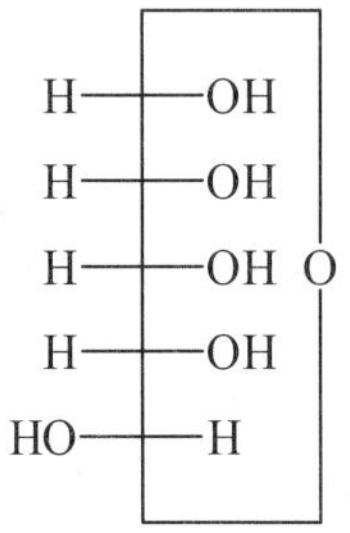

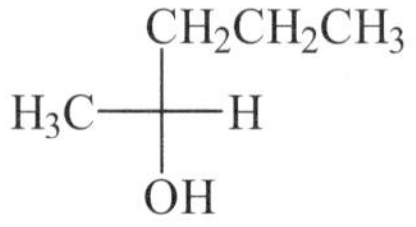

These are not the only possible representations

❑❑ **Convert the following three dimensional structures to Fischer projections.**

a) [3D structure showing CH₃, H, OH, CH₂CH₃ around central carbon]
b) [3D structure showing CH=CH₂, HO, H, phenyl around central carbon]
c) [3D structure showing H₃C, H, Cl on one carbon and H₃C, H, OH on adjacent carbon]

---

a) $H_3C \!-\!|\!-\! CH_2CH_3$ with OH up and H down

b) $H_2C=CH \!-\!|\!-\! $ phenyl, with H up and OH down

c) Fischer projection:
- CH_3
- $Cl - H$
- $HO - H$
- $H - H$
- CH_3

These are NOT the only possible representations

❑❑ **One pure Enantiomer of a new drug is tested and found to have strong anti-cancer activity. Synthesis of this pure Enantiomer is difficult and expensive, while synthesis of the Racemic mixture is quite easy. Why is it not necessarily a good idea to market the drug in racemic form?**

Because the human body contains many chiral compounds, the physiological effects of a pair of Enantiomers may not be identical. The second Enantiomer may have undesirable side effects or may even be toxic. The effects of the second Enantiomer must be studied independently of the original anti-cancer drug.

❑❑ **The enantiomeric excess (ee) of a mixture of Enantiomers is found to be 83.2 % in favor of the (-) isomer. What is the percent optical purity of the sample?**

Enantiomeric excess and percent optical purity are numerically equivalent, therefore the % optical purity = 83.2%.

❑❑ **Which of the following molecules would be optically active?**

a) allene structure with H, CH₃CH₂ on one end and H, CH₃ on the other (C=C=C)
b) cyclohexane with CH₃, CH₃ on one carbon and H, H₃C on another
c) structure with Br, Cl, H on one carbon and Br, H, Cl on adjacent carbon
d) cyclohexene with H₃C and Cl, H substituents

(a) Optically active (remember, a Chiral Center is not always necessary), (b) optically active, (c) optically inactive (meso), (d) optically active.

❑❑ **What are the root words of the designations R and S?**

The designation R is from the Latin word *rectus*, meaning "Right". The designation S comes from the Latin word *sinister*, meaning "Left".

❑❑ **If there is no relationship between the direction of a molecule's rotation of polarized light and its R/S configuration, how is it possible to determine the absolute configuration of a specific molecule?**

The absolute configuration of a molecule can oftern be determined by X-ray crystallography, which essentially takes a picture of the three dimensional positions of the atoms in a crystal. Alternately, a sample of a compound can be synthesized from a molecule whose absolute configuration is already known using a series of reactions that do not disturb the Chiral Centers.

❑❑ **X-ray crystallography is a relatively recent invention. Prior to its use, how did chemists know the absolute configurations of molecules?**

They didn't. In the late 19th century chemists arbitrarily assigned a configuration to (+)-Glyceraldehyde in order to make formulas easier to work with. All other configurations were somehow related to this molecule. In 1951 X-ray experiments showed the assignment to have been correct.

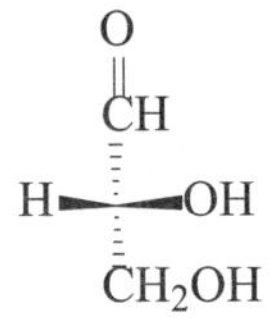

(+)-Glyceraldehyde

❑❑ **What are D and L?**

D and L were stereochemical designations used prior to the R/S convention. (+)-Glyceraldehyde was assigned a specific absolute configuration which was referred to as D. Its Enantiomer, (-)-Glyceraldehyde, was said to have an L configuration. Compounds that had an arrangement of substituents similar to (D)-(+)-Glyceraldehyde were said to have a D configuration and this with an arrangement similar to (L)-(-)-Glyceraldehyde had an L configuration. These old configurations are still used in the names of some compounds, particularly sugars and amino acids.

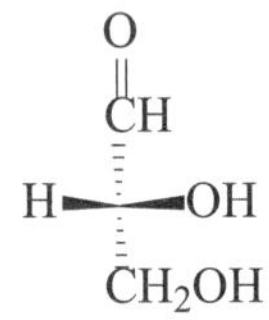

(+)-Glyceraldehyde   (-)-Glyceraldehyde
D configuration   L configuration

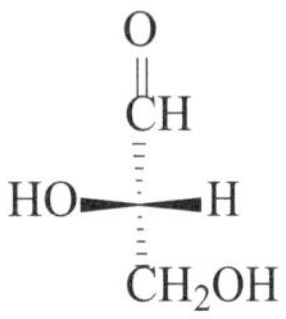

❑❑ **Find all of the chiral Carbons in the following molecules.**

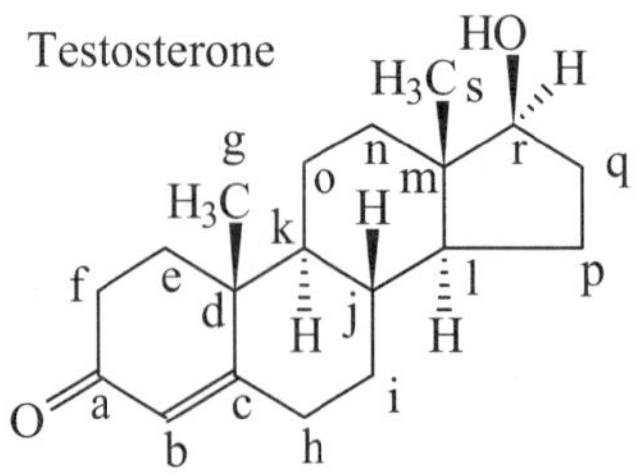

Quinine: Carbons k, l, n, and r are chiral. Testosterone: Carbons d, k, j, l, m, and r are chiral.

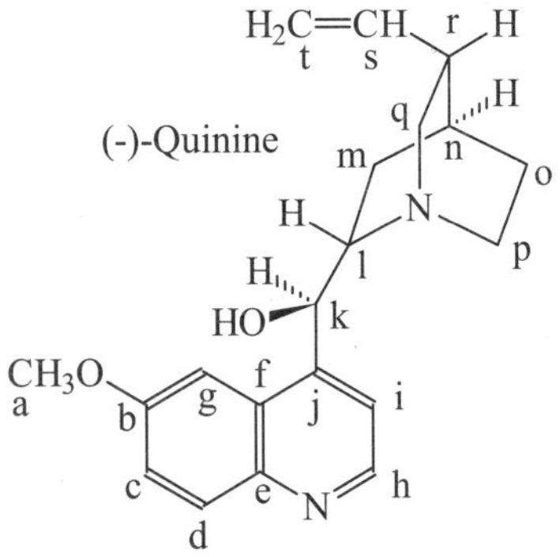

ALKYL HALIDE PEARLS

The great tragedy of science -- the slaying of a beautiful hypothesis by an ugly fact.
Aldous Huxley

☐☐ **What are Alkyl Halides?**

An Alkyl Halide is any compound containing a Halogen atom bonded to an sp^3 hybridized Carbon.

☐☐ **What other types of Organohalogen compounds are there?**

There are two other general classes of organohalides, Aryl Halides and Vinyl Halides. Aryl Halides are compounds which have a Halogen directly bound to an aromatic ring, while Vinyl Halides have the Halogen bonded to the sp^2 hybridized Carbon of a double bond.

☐☐ **What general symbolic formula is often used to represent Alkyl Halides?**

The general formula R-X is often used to represent Alkyl Halides. The R represents an alkyl group, while the X represents a Halogen atom.

☐☐ **The structure of an Alkyl Halide has a great effect on its reactivity. Because of this Alkyl Halides are often classified into four groups. What are they?**

Alkyl Halides are classified as methyl, primary, secondary, or tertiary. The designation refers to the number of alkyl groups attached to the Carbon bearing the Halogen.

$$X-CH_3 \qquad X-CH_2-R \qquad X-\underset{}{\overset{R}{C}}H-R \qquad X-\underset{\underset{R}{|}}{\overset{\overset{R}{|}}{C}}-R$$

Methyl　　　　primary　　　　secondary　　　　tertiary

☐☐ **What is the direction of the polarity of a Carbon Halogen bond?**

Fluorine, Chlorine and Bromine are all more electronegative than Carbon. The electronegativity of Iodine is similar to that of Carbon, but it is much more polarizable due to its large size. As a result, the Carbon Halogen bond is polarized with a slight negative charge on the Halogen atom and a slight positive charge on the Carbon.

$$-\overset{|}{\underset{|}{C}}-X \quad \xrightarrow{}$$

☐☐ **What is the order of Carbon-Halogen bond strengths?**

C-X bond strength decreases as you go down the Halogen family. The C-F bond is the strongest (109 kcal/mol in Fluoromethane), followed by the C-Cl bond (84 kcal/mol in Chloromethane), the C-Br bond (70 kcal/mol), and finally the weakest, the C-I bond (56 kcal/mol).

☐☐ **Is there any correlation between bond strength and bond length in the C-X series?**

Yes, the Carbon-Halogen bond length increases as you go down the Halogen family due to the increasing size of the Halogen atom. This increase in bond length may result in poorer orbital overlap, and hence contribute to the lowering in bond strength.

❏❏ **Fluorine is the most electronegative of the Halogens, yet Chloromethane has the largest dipole moment of the halomethanes. How is this possible?**

The size of a molecule's dipole moment depends primarily on the magnitude of the partial charges and the distance of separation. For the halomethanes, the dipole moment would be expected to increase with the electronegativity of the Halogen and also increase with increasing bond length. Since these two trends run counter to one another, Chloromethane ends up having the largest dipole moment of the series.

❏❏ **What types of forces effect the boiling points of Alkyl Halides?**

The forces between the molecules of a liquid Alkyl Halide consist of a combination of dipole-dipole, dipole-induced dipole, and induced dipole-induced dipole attractions.

❏❏ **Van der Waals forces depend to a large extent on the size of a molecule, yet the comparably sized compounds Ethane and Bromomethane have widely different boiling points (- 89 $^{\circ}$C and 4 $^{\circ}$C, respectively.) Why is this?**

The magnitude of Van der Waals forces is also dependent on the polarizability of a molecule's electrons. The more polarizable the electrons, the easier it is the create an induced dipole. The unshared pairs of electrons on Bromine are much more polarizable than the shared electrons in the C-H sigma bonds of Ethane.

❏❏ **What is meant by the term polarizability?**

Polarizability refers to how easily an atom's or molecule's electron distribution is distorted by a nearby electric field. Tightly held electrons are not very polarizable, whereas loosely held electrons are.

❏❏ **Despite their dipole moments, the boiling points of Alkyl Fluorides are lower than those of Alkanes of comparable molecular weight. How can this be explained?**

Unlike Bromine, Fluorine is very small and holds on to its electrons very tightly. As a result it has very low polarizability and the van der Waals forces are extremely weak - much weaker than in comparable sized Alkanes. The dipole-dipole forces between Alkyl Fluoride molecules can't make up for this.

❏❏ **What are Freons?**

Freons are Chlorofluorocarbons that were manufactured and used in refrigeration systems, as propellants, and as cleaning solvents. They are suspected of causing significant damage to the Earth's ozone layer.

❏❏ **The densities of liquid Alkyl Halides are greater than those of hydrocarbons of comparable molecular weight. Why is this?**

Halogens have a larger mass to volume ratio than similarly sized Alkyl groups. For example, a Bromine atom is about the same size as a methyl group but weighs significantly more (79 amu compared to a Methyl group's 15 amu).

❏❏ **What is the solubility of Alkyl Halides in water?**

All Alkyl Halides are insoluble in water.

❏❏ **What happens when an Alkyl Halide is mixed with water?**

Because of the insolubility of the Alkyl Halides, two layers will form. Monoalkylfluorides and Chlorides are less dense than water and will float on top. Alkylbromides and iodides are denser than water and will sink to the bottom. Polyhalogenation increases the density, so compounds such as CH_2Cl_2 and $CHCl_3$ sink to the bottom.

❏❏ **Arrange the following molecules in order of increasing boiling point:**

a) $CH_3CH_2CH_2Cl$ b) $(CH_3)_2CHCl$ c) $(CH_3)_2CHCH_2Br$

d) $CH_3CH_2CH_2F$ e) $CH_3CH_2CH_2CH_2CH_2Br$

From lowest to highest boiling point: d, b, a, c, e. (remember, increased size = increased boiling point, increased Halogen weight/size = increased boiling point, and increased branching = lower boiling point).

❏❏ **What is the IUPAC system for naming Alkyl Halides?**

First the longest alkyl chain is found (the parent chain.) The parent chain is then numbered from the direction that will give the first substituent encountered (alkyl or Halogen) the lowest number. Halogen substituents are indicated by using the prefixes Fluoro-, Chloro-, Bromo-, and Iodo. The position of each substituent is indicated by using the number of the Carbon to which it is attached. The substituents are listed in alphabetical order at the beginning of the name. If a double bond exists in the parent chain then the parent chain is numbered from the end that gives the lowest numbers to the Carbons of the double bond.

4-chloro-2-methylhexane   (E)-10-bromo-4-decene

❏❏ **How are Alkyl Halides named in the common system of nomenclature?**

The common names of the Alkyl Halides consist of the common name of the Alkyl group followed by the name of the halide as a separate word. They are essentially named as though they were salts.

Butyl Chloride   Isopropyl Iodide   Neopentyl Fluoride

❏❏ **What would the IUPAC and common names be for the following compounds?**

a) [bromocyclohexane] b) [3-fluoropropene structure] c) CH_2Cl_2

(a) Bromocyclohexane and Cyclohexyl Bromide, (b) 3-Fluoropropene and Allyl Fluoride, (c) Dichloromethane and Methylene Chloride

❏❏ **The Trihalogenated Methanes are given different common names than the other Alkyl Halides. What are they called?**

They are known as the Haloforms, the best known of which is Chloroform, $CHCl_3$.

❏❏ **What is the name for saturated Hydrocarbons in which all of the Hydrogens have been replaced by Halogens?**

They are known as Perhaloalkanes.

$CF_3CF_2CF_2CF_2CF_2CF_3$ Perfluorohexane

☐☐ How would you name the following compounds using IUPAC standard nomenclature?

a) [structure with Br, CH3, Br]
b) [cyclohexane with Br and isopropyl]
c) [structure with CH3, CH3, Br]
d) [H2C=CHCF3 structure]

(a) 5,6-Dibromo-2-methyloctane, (b) 1-Bromo-3-isopropylcyclohexane, (c) 2-Bromo-5,6-Dimethylheptane, (d) 3,3,3-Trifluoropropene

☐☐ What would the structures look like for each of the following compounds?

a) (1S, 2S) *trans*-2-Chlorocyclohexanol   b) Iodoform

c) 1-Chloro-2,2-dimethylpropane   d) Isopentylfluoride

a) [cyclohexane with Cl, H (wedge) and OH, H]
b) CHI$_3$
c) Cl–CH$_2$–C(CH$_3$)$_2$–CH$_3$
d) F–CH$_2$CH$_2$CHCH$_3$ with CH$_3$ branch

☐☐ How would you name the following compounds using IUPAC standard nomenclature?

a) [cyclopentane with F, H, H, F]
b) [bicyclic structure with 2 Cl]
c) [cyclohexane with Br, Br]
d) [Cl$_2$C=CCl$_2$]

(a) *trans*-1,2-Difluorocyclopentane, (b) 2,2-Dichlorobicyclo [1.1.1] pentane, (c) *cis*-1,4-Dibromocyclohexane, (d) Tetrachloroethylene

❐❐ **What is necessary to initiate the reaction between Chlorine and a Hydrocarbon?**

Heat or light (visible or ultraviolet).

❐❐ **Is the reaction between Halogens and Hydrocarbons exothermic or endothermic?**

The reaction with Fluorine, Chlorine, and Bromine is exothermic. The reaction with Iodine is endothermic.

❐❐ **What is the order of reactivity of the Halogens in their reaction with Alkanes?**

The order of reactivity is $F_2 \gg Cl_2 > Br_2 > I_2$. Reactions with Fluorine are highly exothermic and difficult to control, reactions with Chlorine and Bromine are synthetically useful, and reaction with Iodine is endothermic and the equilibrium favors reactants rather than products.

❐❐ **What products would you expect from a chlorination of Methane when the reactants are present in a 1:1 ratio?**

If the reactants are mixed in a 1:1 ratio then a mixture of products will result, which will include CH_3Cl, CH_2Cl_2, $CHCl_3$, CCl_4, and leftover Methane.

❐❐ **If one equivalent of Chlorine is used, why isn't Chloromethane the primary product?**

Because addition of a Chlorine to the Methane molecule actually yields a product that is more reactive towards substitution.

❐❐ **What is the best way to obtain only Chloromethane from the chlorination of Methane?**

Use a large excess of Methane relative to Chlorine.

❐❐ **What is the best way to describe the mechanism of the reaction between Halogens and Hydrocarbons?**

They are free radical chain reactions.

❐❐ **What is the mechanism of the reaction between Bromine and Methane to give Bromomethane?**

$$Br-Br \xrightarrow{\text{heat or light}} 2\ Br\cdot \quad \text{Initiation Step}$$

$$CH_3-H + Br\cdot \longrightarrow CH_3\cdot + HBr$$
$$CH_3\cdot + Br-Br \longrightarrow CH_3-Br + Br\cdot$$
Propagation Steps (repeats)

$$H_3C\cdot + \cdot CH_3 \longrightarrow H_3C-CH_3$$
$$Br\cdot + \cdot CH_3 \longrightarrow Br-CH_3$$
$$Br\cdot + \cdot Br \longrightarrow Br-Br$$
Termination Steps

❑❑ **What is a radical?**

A radical is any species that contains an unpaired electron. The unpaired electron is usually symbolized by a single dot next to the formula of the radical species, e.g. $CH_3 \cdot$

❑❑ **What is Homolytic cleavage?**

Cleavage of a two electron covalent bond in such a way that one electron is retained by each of the bonded atoms.

❑❑ **How may Homolytic cleavage be achieved?**

By adding enough energy to a bond to break it. This may be accomplished by heating a molecule or by allowing it to absorb light of the proper frequency, typically ultraviolet.

❑❑ **What other word is also used for Homolytic cleavage?**

It is also referred to as homolysis.

❑❑ **What is meant by Bond Dissociation Energy?**

Bond dissociation energy (BDE) is a measurement of the amount of energy required to cleave a particular bond. There are two types of BDE's that are measured, homolytic bond dissociation energies (the energy to cleave a bond into radicals), and heterolytic bond dissociation energies (the energy to cleave a bond to form a positive and a negative ion).

❑❑ **What would be the expected order of the C-X bond dissociation energies for the monohalomethanes?**

From highest to lowest, CH_3-F, CH_3-Cl, CH_3-Br, CH_3-I.

❑❑ **How reactive are Radical species?**

Radicals are usually extremely reactive due to their incomplete octets.

❑❑ **What structural features effect the stability of Carbon radicals such as the ones found in free radical halogenation?**

Radicals are electron deficient species. Therefore they are stabilized by structural features similar to the ones that stabilize carbocations. Increasing the alkyl substitution on the Carbon bearing the radical increases its stability. So does delocalizing the radical into a π system.

❑❑ **When Propane is monohalogenated, both 1-Halopropanes and 2-Halopropanes are obtained. Would you expect Fluorination, Chlorination, or Bromination to have the higher regioselectivity?**

Bromination would be the most regioselective.

❑❑ **What would be the preferred product in the Monobromination of Propane?**

2-Bromopropane is the preferred product because it goes through the lowest energy (2°) radical.

❑❑ **What are Fishhook arrows?**

When writing mechanisms, fishhook, or single headed, arrows are often used to show the movement of single electrons.

$$Cl-Cl \longrightarrow Cl\cdot + \cdot Cl$$

❏❏ **Using the homolytic BDE data provided, calculate the heat of reaction for the monochlorination of Methane with Cl_2.**

CH_3—H	105 kcal/mol
Cl—Cl	59 kcal/mol
CH_3—Cl	85 kcal/mol
H—Cl	103 kcal/mol

The balanced reaction would be:

$CH_4 + Cl_2 \longrightarrow CH_3Cl + HCl$

On the right hand side one C-H bond and one Cl-Cl bond are being broken. That means that energy must be going in, so we add the BDE's for those two bonds together and the result should have a positive sign:

+ 105 kcal/mol (C-H) + 59 kcal/mol (Cl-Cl) = + 164 kcal/mol = amount of energy needed to break bonds.

On the left hand side a new C-Cl bond and a new H-Cl bond are being formed. Formation of these bonds releases energy, so the BDE's for these bonds are added together with negative signs:

-85 kcal/mol (C-Cl) - 103 kcal/mol (H-Cl) = - 188 kcal/mol = energy released

The overall heat of reaction, ΔH^o, is the sum of the energy consumed and the energy released:

Heat of Reaction = ΔH^o = +164 - 188 = - 24 kcal/mol (exothermic reaction)

❏❏ **In free radical halogenation, what is "chain length"?**

Chain length refers to the number of times the cycle of chain propagation steps repeat. This can be thousands of times.

❏❏ **How many chains begin with each initiation step?**

Two. Each Halogen radical begins its own chain.

❏❏ **How many chains end with each chain termination step?**

Again two. Two radicals from different chains must combine for termination to occur.

❏❏ **How likely is a chain termination step in a radical halogenation chain reaction?**

Not very likely at all. In order for a termination step to occur, two radicals must meet and combine. At any moment the concentration of radicals is actually very low, especially compared to the amount of Alkane and Halogen present. This is why the chain propagation steps can cycle thousands of times before termination happens.

❏❏ **What is a Photochemical reaction?**

A reaction that occurs when a molecule absorbs light energy.

❏❏ **How is a Photochemical reaction indicated when writing a chemical equation?**

Sometimes the word "light" is written over the arrow, but more commonly the symbol "hv" is used, which symbolizes the value of Planck's Constant times the frequency of the light, which is equal to the energy of a photon.

$Br-Br \xrightarrow{h\nu} 2\ Br\cdot$

❏❏ What would you expect the hybridization to be for a methyl radical?

Since there are only three Hydrogens still attached, the hybridization should be sp^2. The radical should be flat, with approximately 120 degree bond angle.

❏❏ What is meant by the phrase "rate-limiting" or "rate-determining" step?

In a multistep reaction, it is the step that is the slowest one. The overall reaction can be no faster than its slowest step.

❏❏ What is the rate-determining step for the free radical halogenation of an Alkane?

The abstraction of a Hydrogen from the Alkane by a Halogen radical.

❏❏ In the monobromination of Propane, what would be the expected statistical distribution of products?

Since there are six Hydrogens total on the end Carbons, and two Hydrogens on the central Carbon, the expected statistical product percentages should be:

$$\frac{2\text{ H}}{8\text{ H (total)}} \times 100 = 25\% \qquad H_3C-\underset{\underset{Br}{|}}{CH}-CH_3$$

$$\frac{6\text{ H}}{8\text{ H (total)}} \times 100 = 75\% \qquad H_3C-CH_2-CH_2-Br$$

❏❏ In reality, bromination of Propane produces 98% 2-Bromopropane and 2% 1-Bromopropane. How is this possible?

In the bromination of Propane the rate-determining step is the abstraction of Hydrogen from the Alkane to give an alkyl radical. There are two competing reactions: abstraction of a Hydrogen form one of the terminal methyl groups, and abstraction of a Hydrogen from the central methylene group. It is the activation energy for each of these reactions that determine their relative rates. Abstraction of a Hydrogen from the methylene produces a secondary radical (lower energy) while abstraction from one of the methyl groups produces a primary radical (higher energy.) Formation of the lower energy secondary radical has much lower activation energy and a much faster rate. Therefore that pathway produces the primary product.

❏❏ Chlorination of Propane produces 55% 2-Chloropropane and 45% 1-Chloropropane. Why doesn't chlorination show as much regioselectivity as bromination?

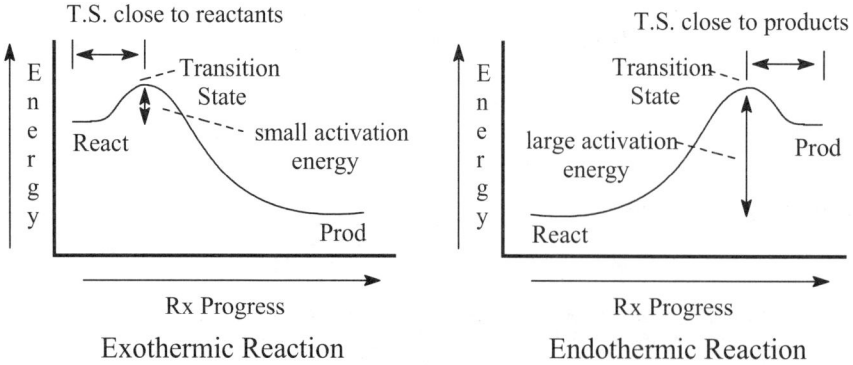

Exothermic Reaction   Endothermic Reaction

The rate determining step (abstraction of a Hydrogen) for chlorination is exothermic, while the analogous step for bromination is endothermic. According to the Hammond Postulate, the transition state for an exothermic reaction looks like the starting materials, while the transition state of an endothermic reaction resembles the products. Since the T.S. for bromination resembles an alkyl radical, whatever structural features that would lower the energy of that radical would also lower the energy of the transition state and decrease the activation energy. Since secondary alkyl radicals are lower energy than primary alkyl radicals, the activation energy for the reaction in which Bromine abstracts a Hydrogen from the central Carbon of Propane is lower, and the rate is therefore much faster. 2-Bromopropane ends up being the preferred product. The transition state for the chlorination reaction, however, resembles the starting materials because the Hydrogen abstraction step in chlorination is exothermic. This means that the transition state looks like a Chlorine radical and a Propane molecule - regardless of whether a primary or secondary Hydrogen is being abstracted. The transition states for the two competing pathways therefore have nearly identical activation energies, so the product distribution for chlorination ends up being about 50:50.

❏❏ **What would you expect the transition states to look like for the rate limiting steps in Bromination and Chlorination of Methane?**

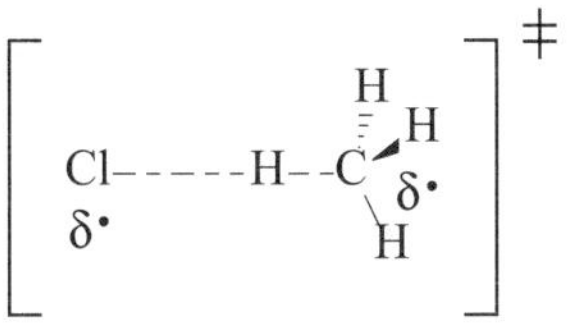

Very small partial radical character on the Carbon. C-H bond slightly weakened. Cl-H bond starting to form

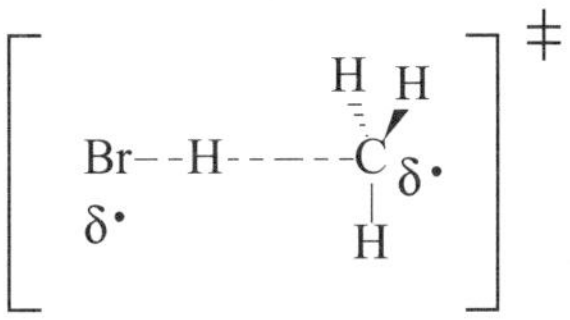

Significant radical character on Carbon. C-H bond almost broken. Br-H bond almost complete. C close to sp2 hybridization.

❏❏ **What does NBS stand for?**

N-Bromosuccinimide.

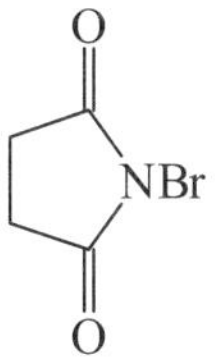

❏❏ **What is NBS typically used for?**

Allylic bromination of Alkenes.

❏❏ **What is the Allylic position in an Alkene?**

The Allylic position refers to the Carbon(s) next to the double bond.

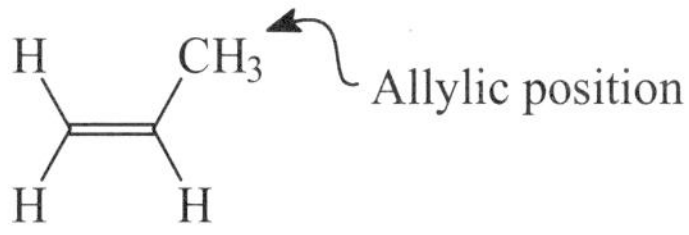

❏❏ **Why isn't Bromine used for Allylic halogenation?**

It can be, but it requires very high temperatures and very low concentrations of Bromine. Higher concentrations of Bromine tend to add to the double bond of Alkenes.

❏❏ **Does NBS react with Alkenes immediately upon mixing?**

No, the reaction is initiated by light or some source of radicals, such as Alkyl Peroxides.

◻◻ **What is the mechanism of Allylic Bromination with NBS?**

Experiments have shown that NBS actually acts as low concentration source of Br_2. Initial reaction of the NBS with light or radical initiators generates as small amount of $Br\cdot$, after which the propagation steps are:

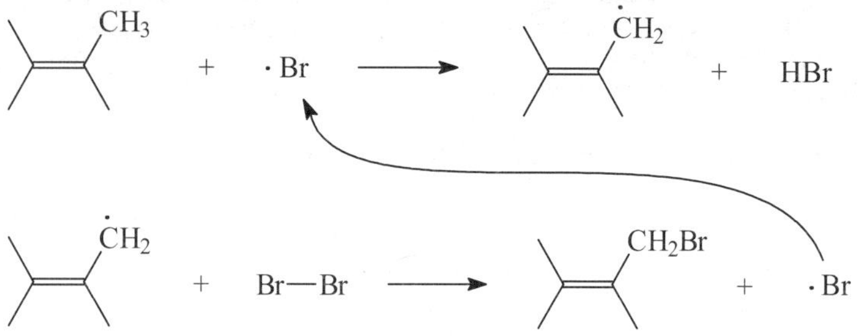

Additional Br_2 is continually generated from the NBS by reaction with the HBr produced in the propagation steps.

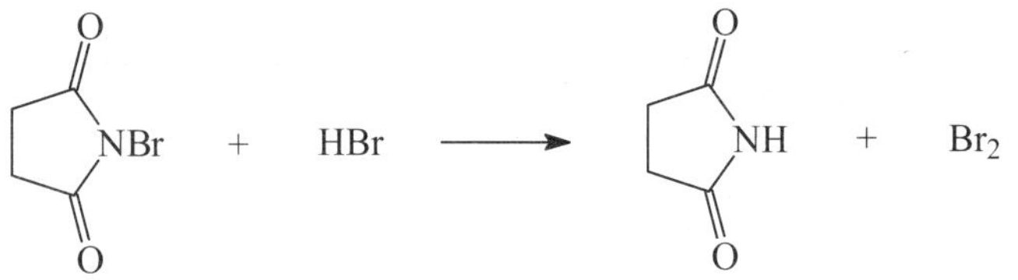

◻◻ **Allylic Bromination with NBS is very regioselective, placing the Bromine in the allylic position virtually 100% of the time, even when other positions are available. Why is this?**

Substitution in the Allylic position is preferred because the Allylic radical formed is more stable than any other radical that would be generated in the substitution at other positions.

◻◻ **Why are Allylic Radicals so stable?**

Because of resonance. There are two resonance forms, which distributes the radical character over a larger area and stabilizes the molecule.

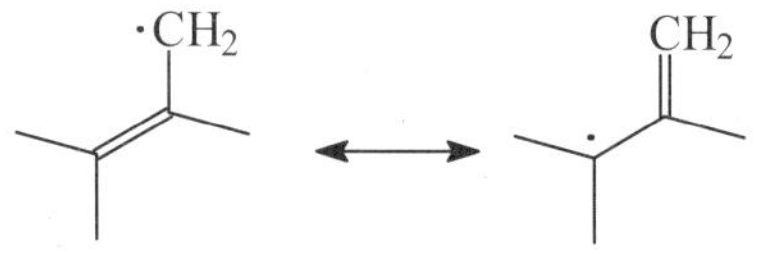

◻◻ **What solvent is usually used for Allylic Bromination using NBS?**

The non-polar solvent Carbon Tetrachloride is usually used.

◻◻ **If Bromine is present during the Allylic Bromination reaction, why doesn't it electrophylically add to the double bond?**

The Bromine is present in very low concentrations. When Br_2 adds to a double bond, one atom of Bromine becomes attached to the Alkene to form the cyclic Bromonium ion. Typically, a Bromide ion formed from another molecule of Br_2 completes the reaction by attacking and opening the cyclic intermediate. Since Bromine is present in such low concentrations, no Bromide is likely to be present at the right time to complete the reaction, and the original addition step simply reverse itself to regenerate the Alkene. Because of this the Allylic Halogenation very successfully competes with electrophylic addition.

❏❏ **What is an organometallic compound?**

An organometallic compound is any compound where a Carbon is directly bound to a metal atom.

❏❏ **What are the products when an Alkyl Halide reacts with two equivalents of lithium metal?**

The products are an Organolithium compound (RLi) and Lithium Halide (LiX).

❏❏ **What is a Gilman Reagent?**

Gilman reagents are Lithium Diorganocopper reagents that result from the treatment of Organolithium compounds with Copper Iodide. They have the general formula R_2CuLi.

❏❏ **What is the purpose of a Gilman Reagent?**

Gilman reagents are used to form new Carbon-Carbon bonds by reaction with Alkyl Chlorides, Bromides, and Iodides. Yields are best when the Alkyl Halides are methyl or primary.

$$R_2CuLi \ + \ CH_3Br \ \xrightarrow{ether} \ RCH_3 \ + \ RCu \ + \ LiBr$$

❏❏ **What is the major drawback of using a Gilman Reagnt?**

Only one of the alkyl groups of the Gilman reagent is transferred. The other is essentially wasted.

❏❏ **What is a Grignard Reagent?**

A Grignard is an Organomagnesium reagent, RMgX.

❏❏ **What can a Grignard Reagent be used for?**

Primarily, Grignard reagents are used for reaction with carbonyl compounds to give a new C-C bond. The final product depends on the nature of the carbonyl compound.

❏❏ **What reaction conditions are necessary to prepare a Grignard Reagent?**

Formation of Grignard reagents requires an Ether as the reaction solvent, and fairly dry conditions (no proton donors).

❏❏ **What functional groups cannot be present in an Alkyl Halide if it is to be used to prepare a Grignard?**

Any functional group that can act as a proton donor (Carboxylic acid, Alcohol, Thiol, primary or secondary Amine, etc.) will result in the decomposition of the Grignard and replacement of the Halogen of the starting Alkyl Halide with a Hydrogen.

❏❏ **How does a Grignard reagent behave chemically?**

The Grignard reagent behaves as though it has a negatively charged Carbon.

$$\overset{\delta-}{CH_3CH_2}-\overset{\delta+}{MgBr}$$

❏❏ **Who discovered the Grignard Reagent?**

Victor Grignard was awarded the 1912 Nobel Prize for Chemistry for his discovery of Organomagnesium compounds and their applications in organic synthesis.

❑❑ **What is the actual structure of the Grignard?**

The actual structure of a Grignard reagent appears to consist of a complex between the reagent and two molecules of Ether.

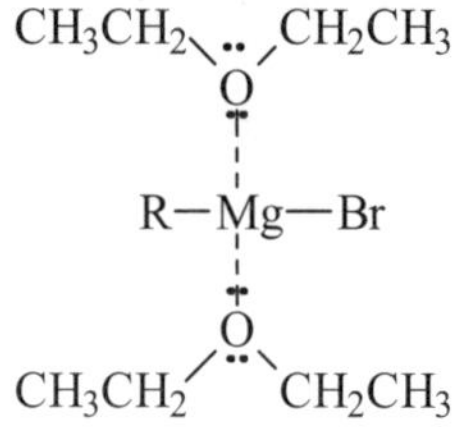

$$\begin{array}{c} CH_3CH_2 \diagdown \!\!\ddot{O}\!\! \diagup CH_2CH_3 \\ | \\ R\!-\!Mg\!-\!Br \\ | \\ CH_3CH_2 \diagup \!\!\ddot{O}\!\! \diagdown CH_2CH_3 \end{array}$$

❑❑ **Using the data provided below, calculate the Percent Ionic Character of the C-Li and C-Mg bonds.**

Element	Electronegativity
C	2.5
Mg	1.2
Li	1.0

$$\text{Percent Ionic Character} = \frac{E(\text{Carbon}) - E(\text{Metal})}{E(\text{Carbon})} \times 100$$

$$\text{C-Li} = \frac{2.5 - 1.0}{2.5} \times 100 = 60$$

$$\text{C-Mg} = \frac{2.5 - 1.2}{2.5} \times 100 = 52$$

❑❑ **What is Nucleophilic Substitution?**

Nucleophilic Substitution is any reaction in which a Nucleophile replaces another group, typically from an sp^3 hybridized Carbon.

❑❑ **What do the designations S_N1 and S_N2 stand for?**

They stand for Substitution Nucleophilic First Order and Substitution Nucleophilic Second order, respectively.

❑❑ **What is a Nucleophile?**

A Nucleophile is an atom or group of atoms that has an unshared pair of electrons that it can donate to form a new covalent bond. Nucleophiles are Lewis bases. The term Nucleophile means "nucleus loving" and refers to the fact that nucleophiles are attracted to centers of partial positive charge.

❑❑ **What is a leaving group?**

In a Nucleophilic Substitution reaction, it is the group that is replaced by the Nucleophile.

❑❑ **What is the substrate?**

In a Nucleophilic Substitution reaction, it is the molecule that is attacked by the Nucleophile.

❏❏ **What is the S$_N$2 mechanism?**

S$_N$2 is a one step mechanism in which a nucleophile attacks a substrate such as an Alkyl Halide and directly kicks out the leaving group:

❏❏ **What is the S$_N$1 mechanism?**

S$_N$1 is a two step mechanism. In the first step, a substrate such as an Alkyl Halide dissociates into a carbocation and a leaving group. In the second step, the carbocation reacts with the nucleophile, which donates a pair of electrons to the carbocation to form a covalent bond. The first step of the reaction is slow and is the rate determining step. The second step is usually very fast.

Step 1   Slow step

Step 2

Attack can occur from either side.

❏❏ **What would be the mechanism of the substitution reaction between Methyl Bromide and Hydroxide ion?**

Methyl Bromide and Hydroxide ion react by an S$_N$2 mechanism.

❏❏ **What would be the mechanism of the substitution reaction between *tert*-Butyl chloride and Methanol?**

This reaction would occur by an S$_N$1 mechanism.

❏❏ **What is the stereochemistry of the S$_N$2 reaction?**

S$_N$2 reactions that occur at Chiral Centers proceed with complete inversion of stereochemistry. The tetrahedral Carbon is turned inside out like an umbrella in a strong wind.

❏❏ **By what other name is this stereochemical result known?**

Walden inversion.

❏❏ **What is the stereochemistry of the S$_N$1 reaction?**

S$_N$1 reactions proceed with almost complete loss of stereochemistry (there is sometimes a very slight preference for inversion). The loss of stereochemistry is due to the fact that the carbocation intermediate is sp^2 hybridized, and therefore flat. The nucleophile can approach form either face, giving a mixture of Enantiomers.

❏❏ **What are the rate equations for the S$_N$1 and S$_N$2 mechanisms?**

S$_N$1: rate = k [substrate]
S$_N$2: rate = k [substrate] [nucleophile]

❑❑ **What structural features of an Alkyl Halide determine whether it will react by S_N1 or S_N2?**

Whether a reaction will proceed by S_N1 or S_N2 is controlled primarily by the amount of alkyl substitution on the Carbon bearing the Halogen. Primary Alkyl Halides tend to react by an S_N2 mechanism, tertiary Alkyl Halides by an S_N1 mechanism, and secondary Alkyl Halides may react by either depending on the specific reaction conditions.

❑❑ **Why do primary Alkyl Halides react by an S_N2 mechanism?**

The S_N2 mechanism requires direct attack by a Nucleophile on the Carbon bearing the Halogen. In a typical primary Alkyl Halide, the approach to this Carbon is clear and the reaction may proceed smoothly. An S_N1 mechanism, however, requires the formation of a carbocation intermediate. For a primary Alkyl Halide, the intermediate would be an extremely unstable primary carbocation. The S_N1 mechanism is therefore unfavorable, and the substitution proceeds by S_N2.

❑❑ **Why do tertiary Alkyl Halides react by an S_N1 mechanism?**

With tertiary Alkyl Halides, the carbocation intermediate formed in the S_N1 mechanism is a reasonably stable tertiary carbocation, so S_N1 can proceed with very little impediment. Direct attack by a nucleophile, however, is quite difficult because of the crowding around the tertiary Carbon. S_N2 is therefore very difficult, and the reaction proceeds by S_N1.

❑❑ **What is the order of stability of carbocations?**

If only saturated carbocations are considered, then the order is $3° > 2° > 1° > CH_3^+$. If benzylic and allylic cations are also considered, then the order is:

$$\begin{bmatrix} 3° \text{ allylic} \\ 3° \text{ benzylic} \end{bmatrix} > \begin{bmatrix} 3° \text{ alkyl} \\ 2° \text{ allylic} \\ 2° \text{ benzylic} \end{bmatrix} > \begin{bmatrix} 2° \text{ alkyl} \\ 1° \text{ allylic} \\ 1° \text{ benzylic} \end{bmatrix} > 1° \text{ alkyl} > CH_3^+$$

❑❑ **What is the easiest way to summarize the competition between S_N1 and S_N2?**

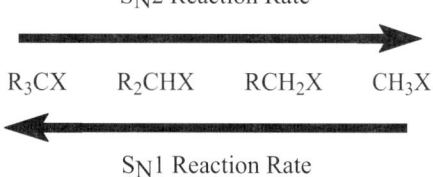

Another good way to summarize the competition is to remember that S_N2 reactions are controlled by steric factors and S_N1 reactions are controlled by electronic factors.

❑❑ **What is a Protic solvent?**

A solvent that capable of forming hydrogen bonds. Solvents that contain -OH groups, for example.

❑❑ **What is an Aprotic solvent?**

A solvent that does not hydrogen bond. Such solvents do not have Hydrogen bonded to an electronegative element.

❑❑ **How do you decide whether a solvent is considered polar or nonpolar?**

This is usually decided based on a solvent's dielectric constant. Solvents are considered polar if their dielectric constant is greater than or equal to 15, nonpolar if it is less than 15.

☐☐ **What type of solvent favors S_N1 and why?**

Polar protic solvents favor the S_N1 reaction. In the rate-determining step of the S_N1 reaction, a carbocation and an anion are usually formed. Anything which stabilizes these two ions will lower the energy of the transition state and increase the rate of the reaction. Polar protic solvents are very capable of solvating both positive and negative ions, thus lowering their energy.

☐☐ **What type of solvent favors S_N2 and why?**

Polar aprotic solvents favor the S_N2 reaction. Because of their structures, polar aprotic solvents solvate cations very well, but anions very poorly. This is important for S_N2, in which the nucleophile is often negatively charged. Since S_N2 is controlled by sterics, anything that impedes access of the nucleophile to the substrate slows down the reaction. If the nucleophile were highly solvated, the cage of solvent molecules would get in the way. Since polar aprotic solvents don't solvate anions, this isn't a problem. Protic solvents, however, would tend to slow down S_N2 reactions.

☐☐ **In the gas phase, the nucleophilicities of the halide ions go in the order $F^- > Cl^- > Br^- > I^-$. In a protic solvent, however, these nucleophilicities are reversed. How is this possible?**

The difference has to do with the amount of solvation of each ion in a polar protic solvent. In the gas phase, the ions are free and there is a direct correlation between the basicity of the ion (ability to donate an electron pair) and its nucleophilicity. F^- is the most basic, so it is the best nucleophile. In a polar protic solvent, the solvation of an anion depends on the concentration of its charge. Small, concentrated charges are the most attractive and are the most highly solvated. Diffuse charges are much less solvated. Consider the two extremes, Fluoride and Iodide. Fluoride is a very small ion with a very strong full negative charge. Such an ion is strongly solvated in protic solvents and develops a large bulky solvent cage. An ion such as iodide, however, has its charge spread out over a much larger area due to the size of the ion. The ion is not as strongly solvated and carries around only a few solvent molecules. Since the ability to act as a Nucleophile depends on the ability to get to and attack the substrate, the Fluoride ion is strongly impeded by solvation and the Iodide is not. This impediment is enough to completely reverse the order of nucleophilicities.

☐☐ **Draw the reaction energy diagram for the S_N2 reaction between Iodoethane and Hydroxide ion.**

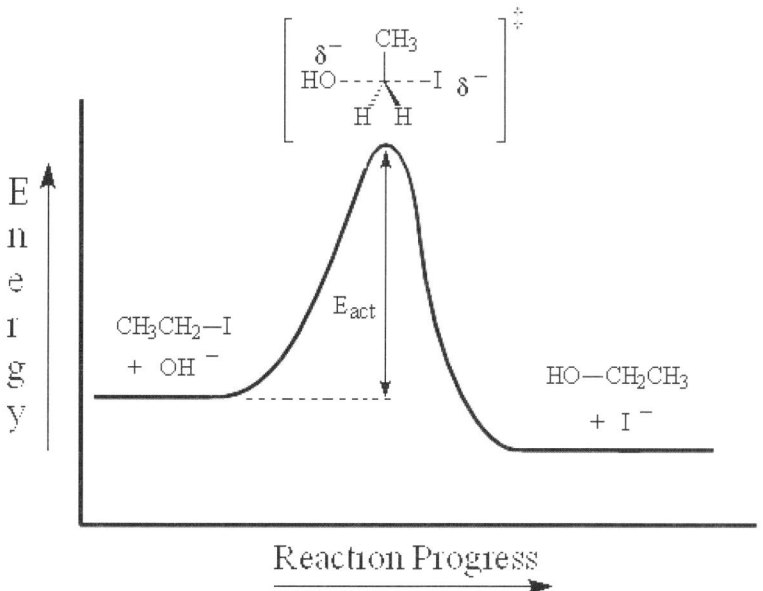

☐☐ **Draw the reaction energy diagram for the S_N1 reaction between 2-Bromo-2-methylbutane and Methanol.**

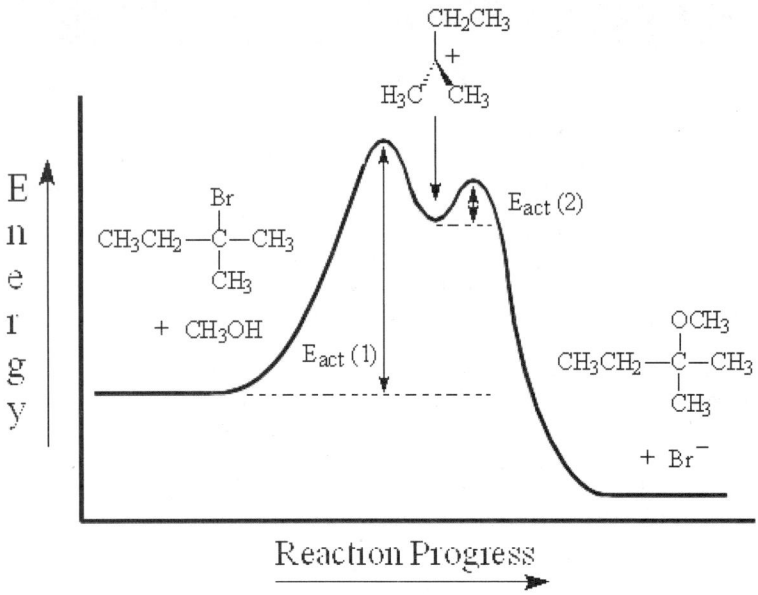

☐☐ **What are the characteristics of a good leaving group?**

Since a partial negative charge is usually placed on a leaving group during the transition state for both S_N1 and S_N2, a good leaving group must be able to stabilize that charge. The best leaving groups are therefore the conjugate bases of strong acids.

☐☐ **Would the following groups be considered good leaving groups or poor leaving groups?**

a) Cl   b) OH   c) NH_2 d) OH_2

(a) Good (conjugate base of HCl), (b) poor (conjugate base of water, a very weak acid), (c) extremely poor (conjugate base of NH_3, and a very strong base too.), (d) good (conjugate base of H_3O^+, the strongest possible acid in water).

☐☐ **What is β-branching, and what effect does it have on the S_N2 reaction?**

In an Alkyl Halide, β-branching is alkyl substitution on the Carbon β to the Halogen (the Carbon next to the one bearing the Halogen). Steric crowding in this position can slow down the rate of an S_N2 reaction even if the Alkyl Halide is primary.

☐☐ **Why is S_N1 not always a good synthetic pathway?**

Because S_N1 reactions go through a carbocation intermediate, there is always a possibility of rearrangement. Also, elimination reactions can often successfully compete with substitution.

☐☐ **What mechanistic pathway is preferred by secondary Alkyl Halides?**

Secondary Alkyl Halides can follow both S_N1 and S_N2 pathways. Which path will be followed depends on the exact reaction conditions, including the choice of the solvent and the identity of the nucleophile. It is even possible for both mechanisms to proceed simultaneously.

☐☐ **Why are Alkyl Fluorides bad choices as substrates for Nucleophilic Substitution?**

The C-F bond is very strong, and therefore requires a large investment of energy to break. In addition, Fluoride ion is not the best leaving group.

❏❏ **Would the following molecules be classified as strong or weak nucleophiles?**

a) Acetic acid b) Methanol
c) Hydroxide Ion d) Iodide

(a) Weak, (b) weak, (c) strong, (d) srong

❏❏ **What is a phase transfer catalyst?**

A phase transfer catalyst is a material that transfers ions back and forth between an aqueous phase and an organic phase.

❏❏ **What are the characteristics of a good phase transfer catalyst?**

A molecule of a phase transfer catalyst must have a hydrophilic portion so that it will dissolve in water and a hydrophobic portion so that it will also dissolve in organic solvents. In addition, it must be capable of forming an ion-pair with the reagent to be carried across the phases.

❏❏ **What are the most typical phase transfer catalysts?**

Tetra-alkylammonium salts.

❏❏ **What is β-elimination?**

β-Elimination is a reaction in which a small molecule such as HCl or H_2O is eliminated from a molecule to form a double bond. The components of the molecule are removed from adjacent Carbons. In Alkyl Halides, the molecule that is eliminated is HX, with the Hydrogen being lost from a Carbon β to the Halogen atom.

❏❏ **What do E1 and E2 stand for?**

E1: Elimination First Order
E2: Elimination Second Order

❏❏ **Draw the mechanism for the E1 reaction.**

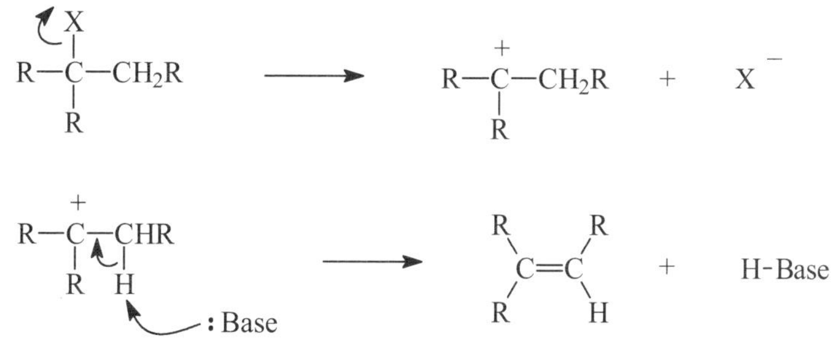

❏❏ **Draw the mechanism for the E2 reaction.**

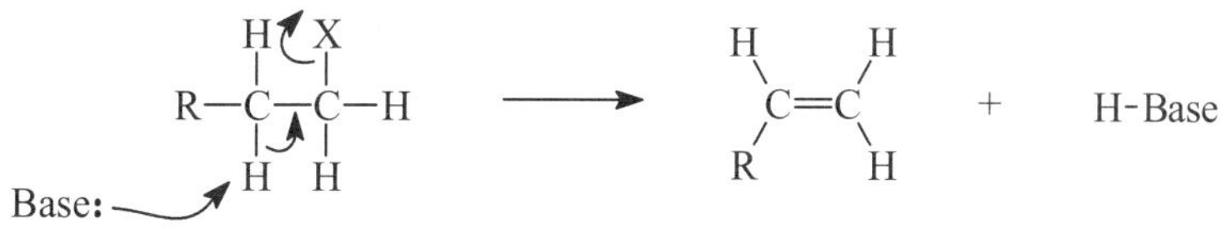

❑❑ **What is the stereoselectivity of the E2 reaction?**

In the E2 reaction, the leaving group and the Hydrogen being extracted are arranged anti-periplanar when the elimination occurs.

❑❑ **What is the stereoselectivity of the E1 reaction?**

E1 elimination is not very stereoselective.

❑❑ **What is the regioselectivity of β-elimination?**

Both E1 and E2 generally follow Zaitsev's Rule; if there is a choice of Hydrogens to lose in the elimination, the one that is preferentially removed is the one that gives the most substituted Alkene. However, in the case of E2 elimination, it must be possible for the Hydrogen being removed to assume an anti periplanar position. If the Hydrogen that would result in the Zaitsev product cannot assume an anti periplanar position, then non-Zaitsev products will be formed.

❑❑ **In the competition between E2 and S_N2, what factors favor each mechanism?**

Branching at the α or β Carbon of the Alkyl Halide decreases the rate of S_N2 (due to steric crowding), and favors the E2 mechanism (due to increased Alkene substitution in the product). Strongly nucleophilic reagents favor substitution, while strongly basic reagents favor elimination. Primary Alkyl Halides that are uncrowded tend to favor substitution, but if bulky bases are used (e.g. Sodium *t*-butoxide) then elimination becomes the major reaction. With secondary Alkyl Halides, bases where the conjugate acid has a pK_a less than or equal to 11 favor substitution, while those whose conjugate acid has a pK_a greater than 11 favor elimination. This is because as the pKa of the conjugate acid increases, basicity increases faster than nucleophilicity.

❑❑ **Why is it so difficult to control the competition between S_N1 and E1?**

Because the rate-determining step is the same for both reactions: loss of the leaving group and formation of a carbocation.

❑❑ **What factors favor each mechanism in the competition between S_N1 and E2?**

E2 is favored by strong bases (hydroxide and alkoxide). S_N1 is favored when poor nucleophiles/bases are used (water, Alcohols, etc.)

ALCOHOL PEARLS

"`You'd better be prepared for the jump into hyperspace. It's unpleasantly like being drunk.'
`What's so unpleasant about being drunk?'
`You ask a glass of water.'"
Douglas Adams in Hitchhikers Guide to the Galaxy

❏❏ **What is the characteristic structural feature of an Alcohol?**

An hydroxyl group (OH) bonded to an sp^3 hybridized Carbon.

❏❏ **What are the bonds angles around the Carbon of Methanol?**

The bond angles are 108.9 degrees.

❏❏ **What is the IUPAC procedure for naming an Alcohol?**

The longest Carbon chain containing the hydroxyl is selected as the parent chain and is numbered from the end closest to the -OH group. The -e suffix of the parent chain's name is changed to -ol and a number is used to show the position of the Alcohol group. Other substituents are named as side chains and numbered appropriately. When the hydroxyl is attached to a Cycloalkane, then the ring is numbered so that the Carbon with the -OH is number 1, unless a different numbering system is required (as in bicyclic ring systems).

❏❏ **How is a compound with multiple OH groups named?**

Instead of using an -ol ending to replace the -e of the parent Alkane name, the -e is retained and the suffix becomes -diol, -triol, etc. to indicate the number of hydroxyls present.

❏❏ **What would the common names be for the following molecules?**

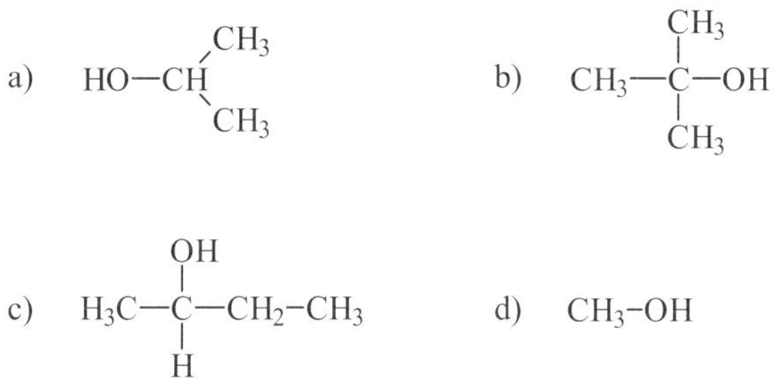

a) HO—CH(CH$_3$)(CH$_3$)   b) CH$_3$—C(CH$_3$)(CH$_3$)—OH

c) H$_3$C—C(OH)(H)—CH$_2$—CH$_3$ d) CH$_3$—OH

(a) Isopropyl Alcohol or Isopropanol, (b) *tert*-Butyl Alcohol or *tert*-Butanol, (c) *sec*-Butyl Alcohol, (d) Methyl Alcohol

❏❏ **What is a compound with a hydroxide bonded to an sp^2 hybridized Carbon called?**

An Enol.

❏❏ **How stable is such a compound?**

Enols generally rearrange spontaneously to give carbonyl compounds. The process is known as tautomerization.

❏❏ **Name the following unsaturated Alcohols**

a) 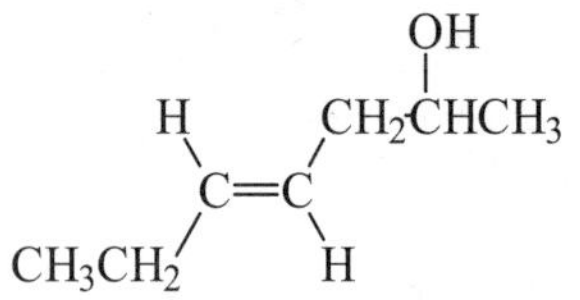 [structure showing CH₃CH₂ and H on one side of C=C, H and CH₂CHCH₃ with OH on the other]

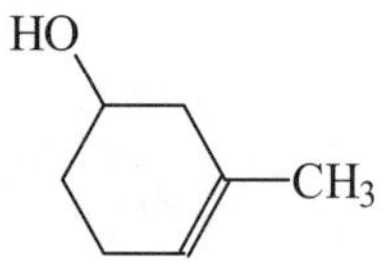

b) [cyclohexene ring with HO and CH₃ substituents]

c) [bicyclic structure with OH]

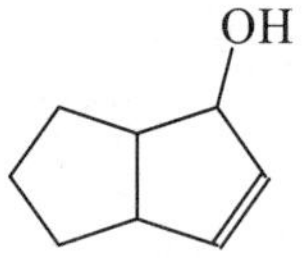

a) (E)-4-Hepten-2-ol, (b) 3-Methyl-3-cyclohexenol, (c) Bicyclo[3.3.0]oct-3-en-2-ol.

❏❏ **How would the following compounds be named using IUPAC standard nomenclature?**

a) 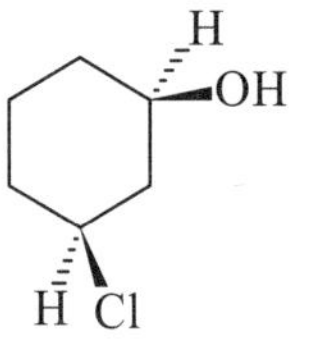 [cyclohexane ring with H, OH, H, Cl substituents]

b) $H_2C=CH-CH_2-OH$

c) 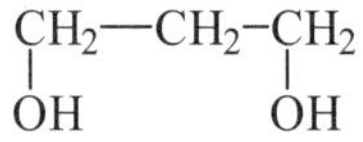 $CH_2-CH_2-CH_2$ with OH groups on end carbons

d) 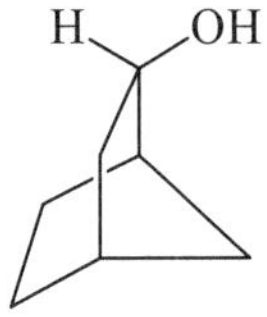 [bicyclic structure with H and OH]

(a) (1R, 2S)-3-Chlorocyclohexanol, (b) 2-Propen-1-ol, (c) 1,3-Propanediol, (d) (2R)-Bicyclo[2.2.1]heptan-2-ol

❏❏ **Due to differences in reactivity, Alcohols are often divided into three classes. What are they?**

Alcohols are generally classified as primary, secondary, and tertiary, referring to the alkyl substitution on the Carbon with the hydroxyl group.

❏❏ **Would the following Alcohols be classified as primary, secondary, or tertiary?**

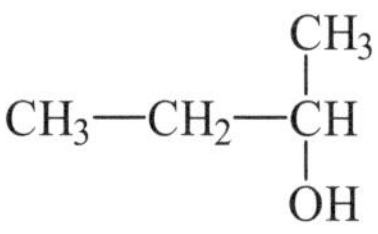

a) 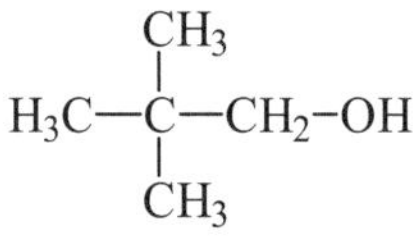 $H_3C-C(CH_3)(CH_3)-CH_2-OH$

b) $CH_3-CH_2-CH(CH_3)-OH$

c) [cyclohexane with HO and CH₃ substituents]

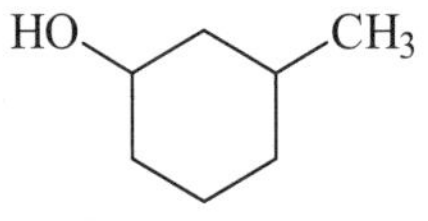

d) 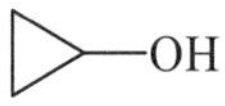 [cyclopropane with OH]

(a) 1°, (b) 2°, (c) 2°, (d) 2°

❏❏ **Are Alcohols considered to be polar or nonpolar compounds?**

All Alcohols have a very polar hydroxyl group, but the measured polarity of the compound depends on its structure. Smaller Alcohols are very polar, while Alcohols with longer hydrocarbon chains tend to be less so.

❑❑ **Ethanol has a boiling point of 78° C, while the isomeric compound Dimethyl Ether has a boiling point of -24° C. Why is this?**

Ethanol is capable of hydrogen bonding, while Dimethyl Ether is not. The strong dipole-dipole attractions between the Ethanol molecules cause the boiling point to be much higher than would normally be expected for a molecule of that size.

❑❑ **For the following compounds, what would be the order of their boiling points?**

a) CH_3CH_2OH

b) $CH_3-\underset{\underset{CH_3}{|}}{\overset{\overset{CH_3}{|}}{C}}-OH$

c) CH_3OH

d) $CH_3CH_2CH_2CH_2OH$

From lowest to highest boiling point: c, a, b, d.

❑❑ **Why is Ethanol infinitely soluble in water while Nonanol is not?**

Small Alcohols are freely soluble in water due to their ability to hydrogen bond. Larger molecules are less soluble because of the increase in the size of the non-polar hydrocarbon portion of their structure.

❑❑ **Arrange the following molecules in order of increasing solubility in water**

a) CH_3CH_2OH

b) $CH_3-\underset{\underset{CH_3}{|}}{\overset{\overset{CH_3}{|}}{C}}-OH$

c) $CH_3\overset{\overset{CH_3}{|}}{C}H_2CH_2CH_2CH_2CH_2OH$

d) $CH_3CH_2CH_2CH_2CH_2OH$

From lowest to highest solubility: c, d, b, a.

❑❑ **What structural features effect the boiling point of an Alcohol?**

The most important structural feature in determining an Alcohol's boiling point is its ability to hydrogen bond. Beyond that, increasing size of the alkyl chain increases the boiling point, as does an increase in the number of hydroxyl groups. Increased branching around the hydroxyl bearing Carbon will lower an Alcohol's boiling point relative to other isomeric Alcohols due to decreased accessibility of the -OH.

❑❑ **What is the typical strength of a hydrogen bond in water?**

The strength of the hydrogen bond between water molecules is about 5 kcal/mol (21 kJ/mol).

❑❑ **Why are hydrogen bonds so strong?**

hydrogen bonding is possible when a proton is bonded to a Fluorine, Oxygen, or Nitrogen. These three elements are extremely electronegative (F = 4.0, O = 3.5, N = 3.0.). Hydrogen only has an electronegativity of only 2.1, but more importantly, when electron density is attracted away from the bonded Hydrogen it leaves behind a very concentrated positive charge. The nucleus of the Hydrogen is not shielded by inner shells of electrons the way that elements of the second period and higher are. The strong dipole that forms makes for very strong dipole-dipole attractions.

❑❑ **What is the solubility of Alcohols in water?**

Smaller Alcohols are totally miscible with water. Solubility decreases as the size of the alkyl portion of the molecule increases. Solubility increases with an increasing number of hydroxyl groups.

❑❑ **What is Ethylene Glycol, and what is it typically used for?**

Ethylene Glycol is also known as 1,2-Ethanediol. It is typically used as a major component in antifreeze.

❑❑ **Alcohols can act as both weak acids and weak bases. How is this possible?**

Alcohols contain both a proton that can be donated (the one attached to the Oxygen) and unshared pairs of electrons that can be donated (on the Oxygen.) Alcohols can therefore act as both Lewis Acids and Lewis Bases.

❑❑ **What is the pK_a of Ethanol?**

The pK_a of ethanol is about 16.

❑❑ **How would you rank the following Alcohols in order of increasing acidity?**

a) $CH_3-\underset{\underset{CH_3}{|}}{\overset{\overset{CH_3}{|}}{C}}-OH$ b) CH_3OH c) CH_3CH_2OH d) $CH_3-\underset{\underset{CH_3}{|}}{CH}-OH$

From least to most acidic: a, d, c, b.

❑❑ ***Tert*-Butyl Alcohol is a much weaker acid in water than Ethanol. Why is this?**

The strength of an acid has a lot to do with the stability of the anion that is formed upon loss of the proton. The more stable the anion, the stronger the acid. Ethanol and *tert*-butanol both form anions with the negative charge on an Oxygen. But in water, solvation of the anion by water molecules can contribute to the ion's stability. The uncrowded ethoxide is much better solvated than the crowded *tert*-butoxide, and is therefore better stabilized. So ethanol ends up being more acidic in water.

❑❑ **Write the equation for the acidity constant of methanol.**

$$K_a = \frac{[CH_3O^-][H_3O^+]}{[CH_3OH]}$$

❑❑ **How do the pK_a's of Alcohols compare to that of water?**

The pK_a of water is 15.7. The pK_a of primary Alcohols is very similar, about 16. Alkyl substitution on the hydroxyl Carbon raises the pK_a, with the pK_a of *tert*-Butanol being around 18.

❑❑ **Under what conditions will Alcohols act as bases?**

In the presence of strong acids.

❑❑ **What is the term applied to protonated Alcohols?**

They are known as alkyloxonium ions.

❑❑ **Write the balanced equation for the reaction between sodium metal and ethanol.**

$$2\ CH_3CH_2OH + 2\ Na \longrightarrow 2\ CH_3CH_2O^-Na^+ + H_2$$

❑❑ **What are salts such as $CH_3O^-Na^+$ usually used for?**

Alkoxides are often used as strong bases in nonaqueous solvents.

❑❑ **How do these salts compare to Sodium Hydroxide in their basicities?**

Alkoxides are stronger bases than hydroxide.

❒❒ **Besides reaction with active metals, how else may an Alcohol be converted to its alkoxide salt?**

Treatment of an Alcohol with any base stronger than alkoxide will convert it to an alkoxide salt. Sodium Hydride (NaH) is often used for this purpose.

❒❒ **Name the following salts.**

a) $CH_3O^- Na^+$ b) $CH_3-\underset{\underset{CH_3}{|}}{\overset{\overset{CH_3}{|}}{C}}-O^- Li^+$ c) $\underset{H_3C}{\overset{H_3C}{}}\!\!\!>\!CH-O^- K^+$

(a) Sodium methoxide, (b) Lithium *t*-butoxide, (c) Potassium isopropoxide

❒❒ **Why is the reaction between an Alcohol and an alkali metal irreversible?**

In addition to producing large quantities of energy, the reaction produces Hydrogen, which is released as a gas.

❒❒ **What are the most common reagents used for converting an Alcohol to an alkyl chloride?**

HCl, PCl_5, and $SOCl_2$.

❒❒ **What type of Alcohol is best for reaction with a Hydrogen Halide?**

Tertiary Alcohols react readily with Hydrogen Halides at room temperature.

❒❒ **How are these reactions usually carried out?**

Water-soluble Alcohols are dissolved in concentrated Hydrogen Halide solutions and the resulting Alkyl Halide forms an insoluble layer. Water insoluble Alcohols are treated with gaseous HX, which is bubbled through a solution of the Alcohol dissolved in Diethyl Ether or THF.

❒❒ **What reaction conditions are necessary for conversion of primary and secondary Alcohols?**

Primary and secondary Alcohols do not react with HCl at reasonable rates. Treatment with HBr and HI at reflux temperatures will convert them to Alkyl Halides, but rearrangement of secondary Alcohols sometimes occurs.

❒❒ **When 3-Pentanol is reacted with concentrated HBr under reflux conditions, the major product is 3-Bromopentane, but some 2-Bromopentane is also obtained. How could this happen?**

This product is the result of a rearrangement of the secondary carbocation produced in the reaction.

☐☐ **What is the mechanism of the reaction between Alcohols and Hydrogen Halides?**

$$R-\underset{R}{\overset{R}{C}}-\ddot{O}H \;+\; H-X \longrightarrow R-\underset{R}{\overset{R}{C}}-\overset{H}{\underset{}{\overset{+}{O}}}-H \;+\; X^-$$

$$R-\underset{R}{\overset{R}{C}}-\overset{H}{\underset{}{\overset{+}{O}}}-H \longrightarrow R-\underset{R}{\overset{R}{C}}{}^+ \;+\; H_2O$$

$$R-\underset{R}{\overset{R}{C}}{}^+ \;+\; X^- \longrightarrow R-\underset{R}{\overset{R}{C}}-X$$

The mechanism is a type of S_N1 mechanism, in which the hydroxide is first protonated to convert it to a good leaving group. The molecule then dissociates to form a carbocation in the rate-determining step, and subsequently combines with a halide ion to form the final product.

☐☐ **A better method for converting primary and secondary Alcohols to alkyl bromides is reaction with Phosphorus Tribromide. Why would this be better than using HX?**

The method works well for primary and secondary Alcohols, is carried out under milder conditions (0° C rather than reflux), and results in much less rearrangement with secondary Alcohols.

☐☐ **What type of mechanism is followed in the reaction between primary or secondary Alcohols and Phosphorus Tribromide?**

The mechanism is essentially S_N2. The substitution step is proceeded by conversion of the Alcohol to a protonated dibromophosphite, which makes the -OH a good leaving group. This group is then displaced by a bromide ion.

$$R-CH_2-\ddot{O}H \;+\; \underset{Br}{\overset{Br}{\underset{|}{P}}}-Br \longrightarrow R-CH_2-\underset{H}{\overset{+}{\ddot{O}}}-PBr_2 \;+\; Br^-$$

$$Br^- \;+\; R-CH_2-\underset{H}{\overset{+}{\ddot{O}}}-PBr_2 \xrightarrow{S_N2} R-CH_2-Br \;+\; H\ddot{O}-PBr_2$$

☐☐ **When Chiral Alcohols react with Thionyl Chloride, there are two possible stereochemical outcomes: retention and inversion. What are the reaction conditions necessary to produce each?**

If the reaction is run in the presence of Pyridine or a tertiary Amine, the result is inversion. If the reaction is run in Ether with no Amine present, the result is retention.

☐☐ **Why do these similar reactions have such different stereochemical results?**

Because the mechanisms are different. The first steps for each reaction are the same, initial reaction of the Alcohol with Thionyl Chloride to give an Alkyl Chlorosulfite:

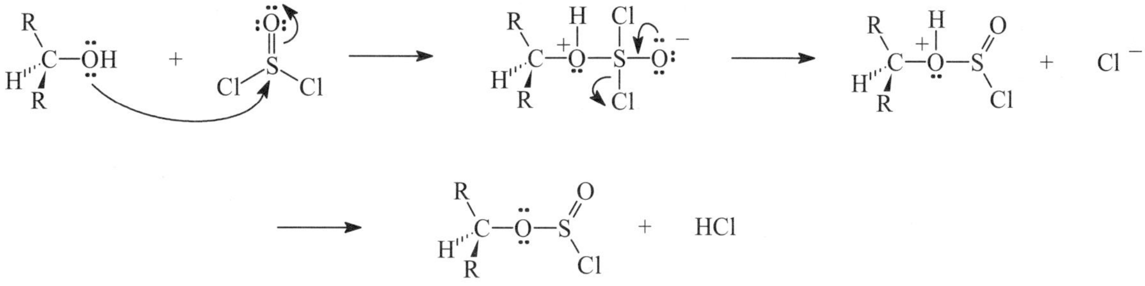

An Alkyl Chlorosulfite

When Pyridine is present, the HCl produced in the reaction protonates the Nitrogen, and the chloride ion is freed. The chloride then attacks the Alkyl Chlorosulfite from the rear (S_N2) to give inversion of configuration.

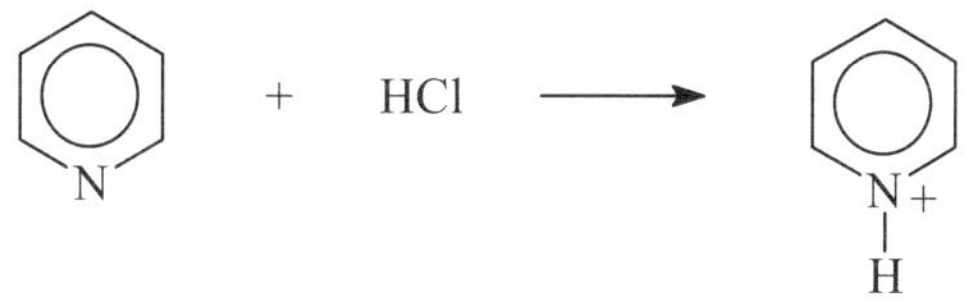

Inversion

When no Pyridine is present, there is no nucleophile to attack the Alky Chlorosulfite. The intermediate dissociates to give a carbocation, SO_2, and chloride. In the relatively nonpolar Ether solvent, the ions form a tight ion pair surrounded by a cage of solvent molecules. The chloride produced by the dissociation attacks the carbocation on the same side from which it was produced, resulting in retention of configuration. This mechanism is known as S_Ni (substitution nucleophilic internal).

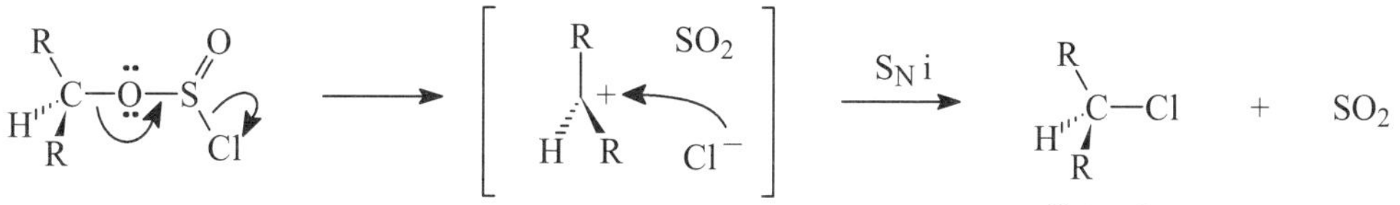

Retention

☐☐ **What is an Alkyl Sulfonate?**

An alkyl Sulfonate is the product of the reaction between an Alcohol and a Sulfonyl Chloride.

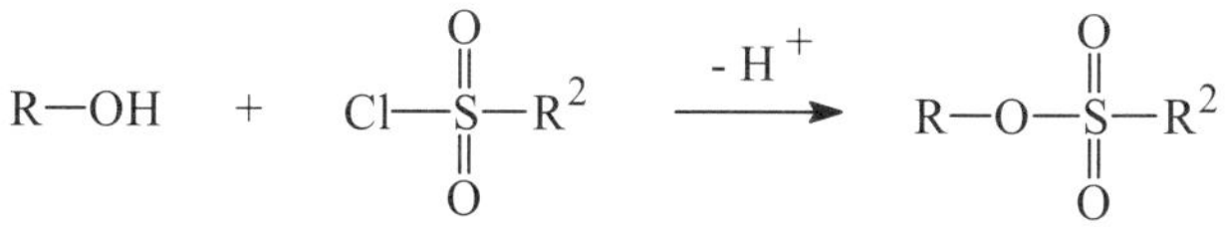

An Alkyl Sulfonate

❏❏ **Why are Alkyl Sulfonates usually made?**

To convert a hydroxyl group into a better leaving group for substitution.

❏❏ **Draw the structures of Tosyl Chloride and Mesyl Chloride.**

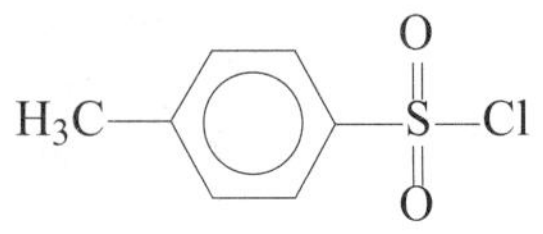

$$H_3C\text{-}C_6H_4\text{-}SO_2\text{-}Cl \qquad\qquad H_3C\text{-}SO_2\text{-}Cl$$

p-Toluenesulfonyl Chloride, also known as Tosyl Chloride or Ts-Cl

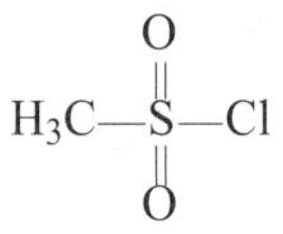

Methanesulfonyl Chloride, also known as Mesyl Chloride or Ms-Cl

❏❏ **What is the stereochemical result of converting an Alcohol to an Alkyl Sulfonate?**

Retention. No bonds to the Chiral Center are broken.

❏❏ **What is a Pinacol rearrangement?**

When 1,2-Diols (glycols) are dehydrated with acid, the product is not an Alkene, but a carbonyl compound. Formation of this compound proceeds through a rearrangement in which an alkyl group or a hydride migrates.

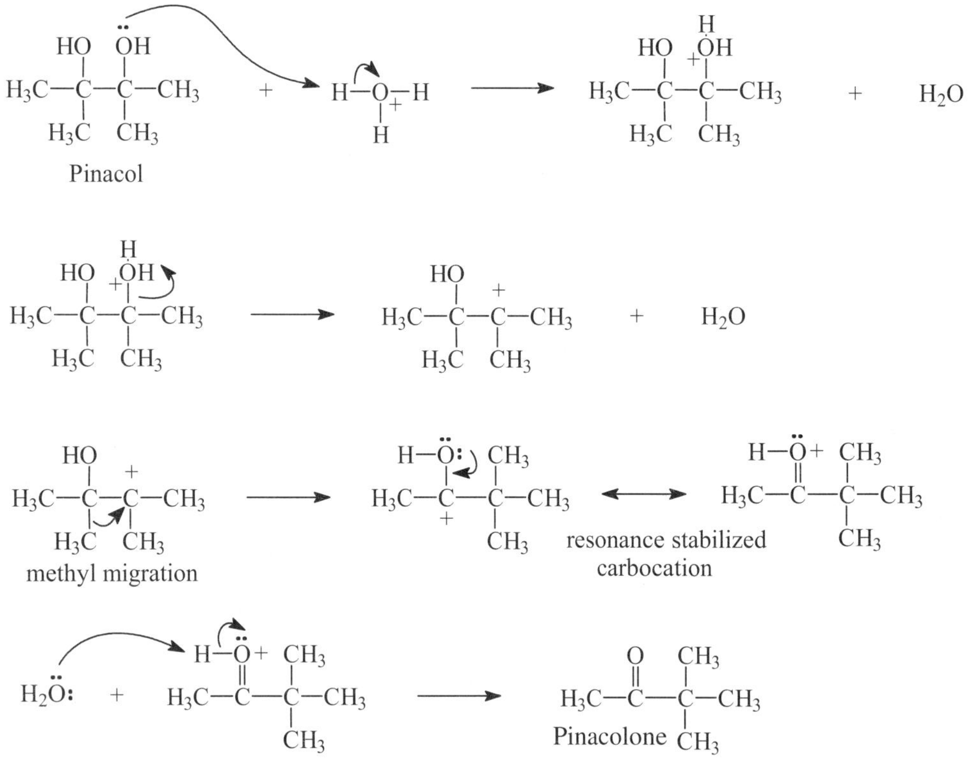

[Reaction scheme: Pinacol + H$_3$O$^+$ → protonated diol + H$_2$O → carbocation after loss of water → methyl migration → resonance stabilized carbocation → deprotonation by H$_2$O → Pinacolone]

❏❏ **What would be the expected primary product(s) of each of the following reactions?**

a) $\text{H}_3\text{C}-\text{CH}=\text{CH}_2$ (with H on the same carbon as CH₃) + H_2O $\xrightarrow{\text{H}_2\text{SO}_4}$

b) 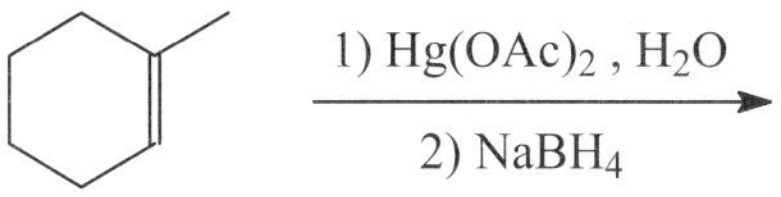 1-methylcyclohexene $\xrightarrow[\text{2) NaBH}_4]{\text{1) Hg(OAc)}_2, \text{H}_2\text{O}}$

c) 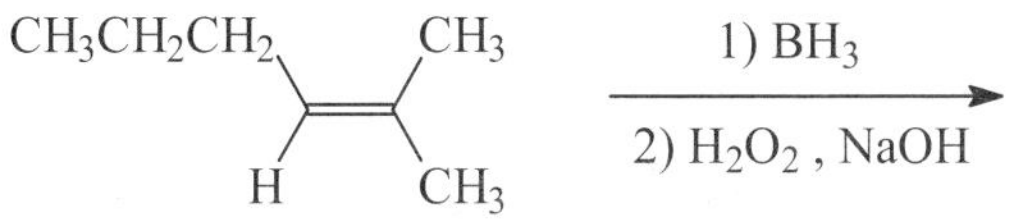 $\text{CH}_3\text{CH}_2\text{CH}_2\text{C(CH}_3\text{)=CH(CH}_3\text{)}$ (cis with H and CH₃) $\xrightarrow[\text{2) H}_2\text{O}_2, \text{NaOH}]{\text{1) BH}_3}$

d) $\text{CH}_3\text{CH}_2\text{CH}_2\text{OH}$ + Na $\longrightarrow$

---

a) $\text{H}_3\text{C}-\overset{\text{OH}}{\underset{}{\text{CH}}}-\text{CH}_3$

b) 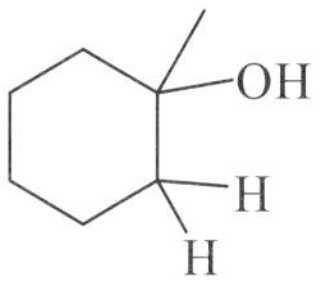 1-methylcyclohexanol (OH and CH₃ on same carbon, H on adjacent carbon)

c) 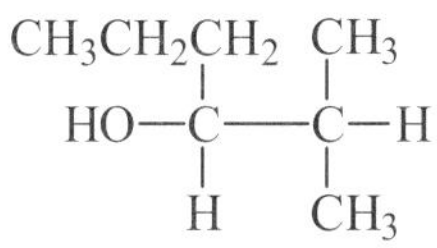 $\text{CH}_3\text{CH}_2\text{CH}_2-\overset{\text{H}}{\underset{\text{HO}}{\text{C}}}-\overset{\text{CH}_3}{\underset{\text{CH}_3}{\text{C}}}-\text{H}$

d) $\text{CH}_3\text{CH}_2\text{CH}_2\text{O}^-\ \text{Na}^+$ + H_2

❏❏ **What is the primary source of Ethyl Alcohol?**

Fermentation of carbohydrates.

❏❏ **When Ethanol from fermentation is distilled, pure ethanol is not obtained. Instead, a mixture of 95% ethanol and 5% water distills. Why is this?**

Ethanol and water form an azeotrope (95% ethanol and 5% water) that distills at a temperature lower than that of pure Ethanol. As a result, the azeotrope distills first, and continues to distill until either the water or the Ethanol is exhausted.

❏❏ **What is denatured Alcohol?**

Since Alcoholic beverages are usually taxed, Ethanol for use in the lab is often contaminated with toxic materials such as Methanol or Isopropanol so that it cannot be consumed. Ethanol treated in such a manner is not taxed as a liquor.

❏❏ **When Ethanol is metabolized by the liver, what is the first product formed?**

The Ethanol is oxidized to Acetaldehyde.

☐☐ **Ethanol, though not precisely nontoxic, may be used as a beverage. The closely related Alcohol Methanol, however, is extremely toxic, often causing blindness and death on ingestion. Why is there such a difference?**

The primary difference is not in the Alcohols themselves, but in the oxidation products formed in the liver. Ethanol is converted to Acetaldehyde, which is then further oxidized to acetate ion. Methanol is instead converted to Formaldehyde, which is highly toxic.

☐☐ **Grignard reagents react with carbonyl compounds to give Alcohols. What types of carbonyl compound would be required to produce primary, secondary, and tertiary Alcohols, respectively?**

To produce a primary Alcohol, the Grignard would have to be reacted with Formaldehyde. To obtain a secondary Alcohol, an Aldehyde would be used, and to get a tertiary Alcohol would require reaction with a Ketone.

☐☐ **What are the typical conditions used to bring about the dehydration of an Alcohol?**

The Alcohol is generally heated with 85% Phosphoric Acid or concentrated Sulfuric Acid.

☐☐ **What would be the expected major product if the following Alcohols were dehydrated?**

a) $CH_3CH_2-CH(Cl)-CH_3$ b) $CH_3CH(CH_3)-C(Br)(CH_3)-CH_2CH_3$ c) cyclohexane with Cl, and two CH$_3$ groups

a) $CH_3CH=CH-CH_3$ b) $CH_3-C(CH_3)=C(CH_3)-CH_2CH_3$ c) cyclohexene with two CH$_3$ groups

☐☐ **What is the order of reactivity of Alcohols in an acid catalyzed dehydration reaction?**

The order of reactivity is $3° > 2° > 1°$. Tertiary Alcohols react readily at temperatures slightly above room temperature, secondary Alcohols require higher temperatures, and primary Alcohols require Sulfuric Acid and temperatures of up to 180 °C.

☐☐ **What kind of mechanism does acid catalyzed dehydration of Alcohols generally follow?**

Dehydration with acid generally follows an E1 mechanism, though primary Alcohols with little ☐-branching probably follow an E2 mechanism.

☐☐ **What synthetic drawback does acid catalyzed dehydration of Alcohols possess?**

Acid catalyzed dehydration, especially of primary and secondary Alcohols, is often accompanied by rearrangement.

☐☐ **Alkenes can be converted to Alcohols by an acid catalyzed hydration reaction whose mechanism is the exact opposite of the acid catalyzed dehydration of Alcohols. What determines which reaction will occur?**

There is actually equilibrium between the dehydrated form (Alkene) and the hydrated form (Alcohol) of the molecule. As with all equilibria, it follows Le Chatelier's principle: The presence of large quantities of water favors formation of the Alcohol, while the presence of only small quantities of water (or the removal of water) favors the Alkene. If concentrated acids are used and the temperature is raised above 100 °C to boil off any water, then the equilibrium is shifted toward dehydration. If dilute acids are used, then the water present shifts the equilibrium toward the Alcohol.

❏❏ **2-Propanol reacts in the presence of strong acid to give Propene. What is the mechanism of this reaction?**

[Mechanism diagram: protonation of 2-propanol hydroxyl by hydronium ion to give protonated alcohol and water; loss of water to form isopropyl carbocation; deprotonation by water to give propene and hydronium ion.]

❏❏ **In the acid catalyzed dehydration of 2-Propanol, what is the rate determining step?**

Formation of the carbocation.

❏❏ **What is the Lucas Reagent?**

The Lucas Reagent is anhydrous Zinc Chloride dissolved in concentrated HCl. It is used to convert low molecular weight primary and secondary Alcohols into chloroalkanes.

❏❏ **1-Pentanol reacts with the Lucas Reagent (at reflux) to give high yields of 1-chlorobutane. The isomeric Alcohol 2,2-Dimethyl-1-propanol, however, gives 2-Chloro-2-methylbutane as the primary product. Why is this?**

1-Pentanol reacts by an S_N2 mechanism, reacting cleanly and giving a high yield of product. 2,2-Dimethyl-1-propanol, though primary, is too crowded to go through an S_N2 mechanism due to β-branching. As a result, the reaction goes through an unstable carbocation intermediate, which rearranges before it gives the final product.

❏❏ **What is the Fischer esterification reaction?**

Fischer esterification is an acid catalyzed reaction between a Carboxylic Acid and an Alcohol to yield an Ester and water.

❏❏ **What is the reagent of choice for converting a primary Alcohol to an Aldehyde?**

Pyridinium Chlorochromate (PCC, $C_5H_5NH^+ ClCrO_3^-$).

❏❏ **Why are tertiary Alcohols typically immune to oxidation?**

Tertiary Alcohols have no Hydrogen on their hydroxyl-bearing Carbon, so there is no way to remove the components of H_2 to form a Carbon-Oxygen double bond.

❏❏ **What happens when oxidation of a tertiary Alcohol is attempted?**

Under normal conditions no reaction will take place. With strong oxidizers and high temperatures, C-C bond cleavage yields a complex mixture of products.

❏❏ **An unlabelled bottle of Alcohol is suspected of being one of the following:**

a) 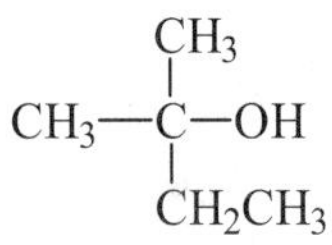 CH₃—C(CH₃)(CH₂CH₃)—OH   b) CH₃CH₂CH₂—C(OH)(H)—CH₃   c) 1-methylcyclohexan-1-ol (H₃C, OH on cyclohexane)

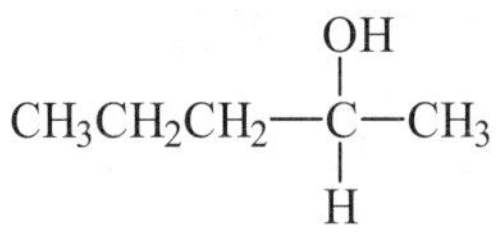

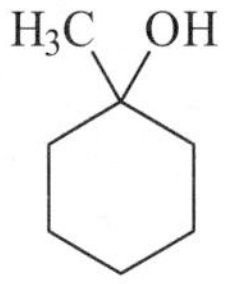

Treatment with Jones' Reagent (Chromic Acid in Sulfuric Acid) produces a green color upon addition. What is the likely identity of the Alcohol?

The Alcohol is compound (b). Tertiary Alcohols such as (a) and (c) will not react with Chromic Acid.

❏❏ **What kind of intermediate is formed in oxidation of an Alcohol with Chromic Acid?**

A Chromate Ester.

❏❏ **An unknown organic compound was treated first with Osmium Tetroxide, and then the product of that reaction was reacted with Periodic Acid. The sole organic material isolated at the end is Acetaldehyde. What was the starting unknown?**

Periodic acid is a reagent used to cleave diols to give carbonyl compounds. If the product was Acetaldehyde, then the diol must have been 2,3-Butanediol. Osmium Tetroxide is used to synthesize diols from Alkenes. To make 2,3-Butanediol, the starting material must have been *cis* or *trans*-2-Butene. There is no way to tell from the data which stereoisomer was used.

❏❏ **What would be the primary organic product obtained in each of the following reactions?**

a) 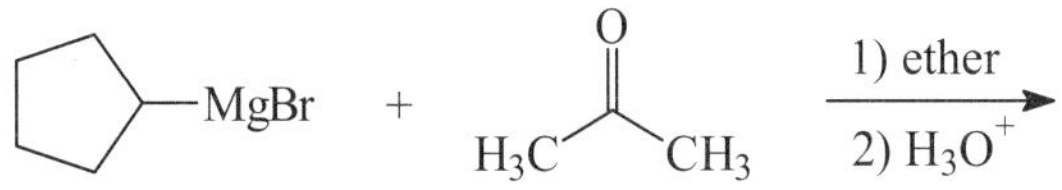 cyclopentyl—MgBr + (CH₃)₂C=O → 1) ether 2) H₃O⁺

b) 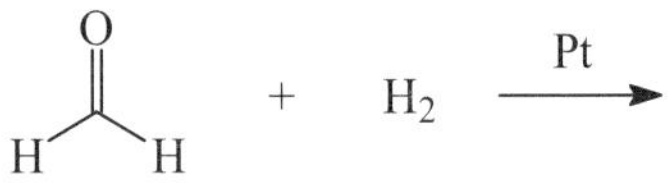 H—C(=O)—H + H₂ → Pt

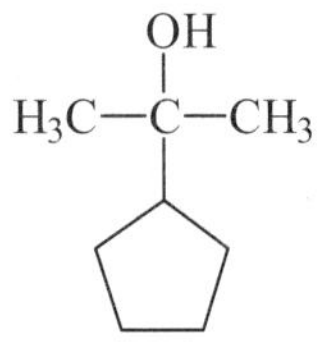

c) CH₃CH₂—C(=O)—CH₃ → NaBH₄ / Methanol

a) H₃C—C(OH)(CH₃)—cyclopentyl   c) CH₃OH   d) 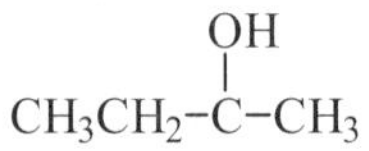 CH₃CH₂—C(OH)(H)—CH₃

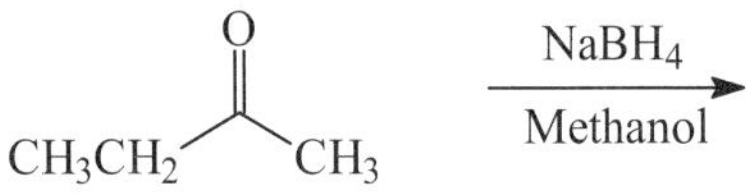

❏❏ **What reagent could be used to convert 1-Propanol into each of the following compounds?**

a) 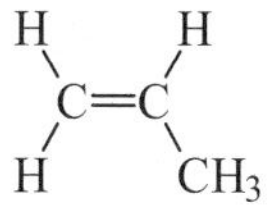 H₂C=CHCH₃ (as drawn: H,H on one carbon; H,CH₃ on other)   b) 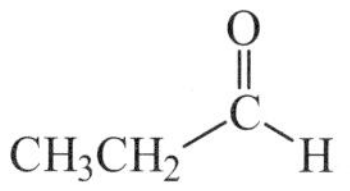 CH₃CH₂C(=O)H   c) 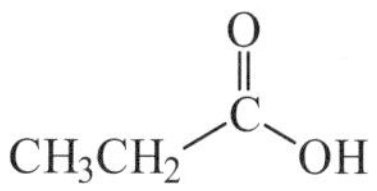 CH₃CH₂C(=O)OH   d) CH₃CH₂CH₂O⁻ Li⁺

(a) Sulfuric Acid, (b) PCC, (c) Chromic Acid, Lithium metal

❏❏ **Alcohols may sometimes be prepared from Alkyl Halides by treatment with hydroxide ion. Why is this approach rarely used for synthesis?**

Two reasons: (1) Only substrates that cannot undergo E2 are useful for the reaction, and (2) Most Alkyl Halides are prepared from Alcohols because the corresponding Alcohols are more readily available.

ETHER AND EPOXIDE PEARLS

*The room contained an odd assortment of odors, of which the
most emphatic at the moment seemed to be the pungent aftermath
of cordite and the sickish aroma of ether.*
Raymond Chandler

❑❑ **What is an Ether?**

An Ether is a compound in which Oxygen is bonded to two Carbon atoms: C-O-C.

❑❑ **How do Ethers and Epoxides differ?**

Ethers and Epoxides differ strongly in their chemical reactivity. Epoxides are much more reactive than typical Ethers.

❑❑ **By what other name are Epoxides known?**

The official IUPAC name for this class of compounds is the Oxiranes.

❑❑ **How are Ethers named in IUPAC nomenclature?**

The longest Carbon chain is chosen as the parent compound and the smaller -OR substituent is named as an alkoxy substituent.

❑❑ **What is the method for forming the common names of Ethers?**

The alkyl groups attached to the Oxygen are listed in alphabetical order and the word Ether is then added. If the Ether is symmetric, then the prefix di- is used before the name of the alkyl group.

❑❑ **How would you name the following compounds using IUPAC standard nomenclature?**

a) $CH_3CH_2-O-CH_3CH_2OH$

b) 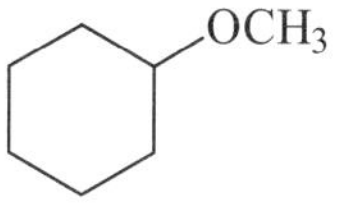 cyclohexane-OCH_3

c) $CH_3CH_2CH_2-O-CH_2CH_2CH_2CH_3$

d) 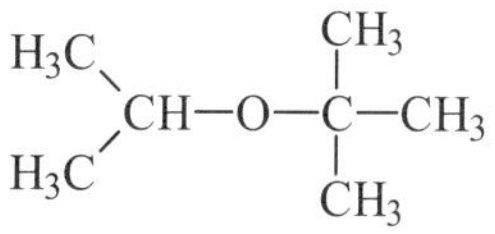 $(H_3C)_2CH-O-C(CH_3)_3$

(a) 2-Ethoxyethanol, (b) Methoxycyclohexane, (c) 1- Propoxy butane, (d) 2-Isopropoxy-2-methylpropane

❑❑ **What compound is known simply as "Ether"?**

Diethyl Ether: $CH_3CH_2OCH_2CH_3$.

❑❑ **What is the classic (non-chemical) use of "Ether"?**

As an anesthetic.

❏❏ **Draw the structures of the following compounds using their common names as a guide.**

a) Dibutyl Ether
b) Isobutyl Methyl Ether
c) Cyclopentyl Ethyl Ether
d) *Tert*-Butyl Butyl Ether

a) $CH_3CH_2CH_2CH_2-O-CH_2CH_2CH_2CH_3$

b) 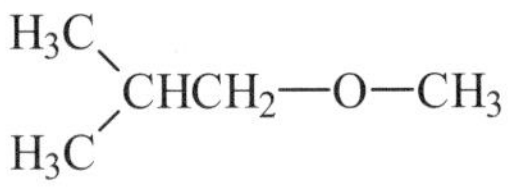 $(H_3C)_2CHCH_2-O-CH_3$

c) 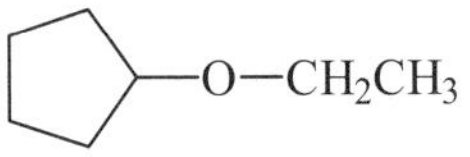 cyclopentyl$-O-CH_2CH_3$

d) 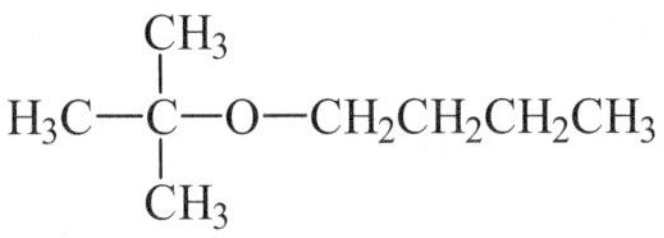 $H_3C-C(CH_3)_2-O-CH_2CH_2CH_2CH_3$

❏❏ **What is Diglyme?**

Diglyme stands for <u>Di</u>ethylene <u>Gly</u>col Di<u>me</u>thyl <u>E</u>ther, a common solvent.

$H_3C-O-CH_2CH_2-O-CH_2CH_2-O-CH_3$

❏❏ **What are the IUPAC names of the following cyclic Ethers?**

a) 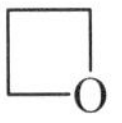 oxetane ring
b) 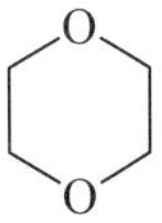 1,4-dioxane ring
c) 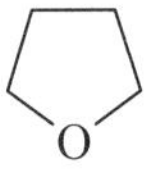 oxolane ring
d) 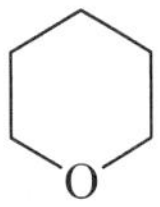 oxane ring

(a) Oxetane, (b) 1,4-Dioxane, (c) Oxolane, (d) Oxane.

❏❏ **C and D above are better known by their common names. What are they?**

Compound C is better known as Tetrahydrofuran (THF), and compound D is known as Tetrahydropyran (THP).

❏❏ **What is the boiling point of the most common Ether, Diethyl Ether?**

About 35 °C.

❏❏ **How do the boiling points of Ethers compare with those of similar hydrocarbons?**

The boiling points of Ethers are very close to those of hydrocarbons with similar molecular weight. Consider Diethyl Ether, (MW = 74, BP = 35 °C) and Pentane (MW = 72, BP = 35 °C).

❏❏ **Why do Ethers boil so much lower than isomeric Alcohols?**

Alcohols are capable of hydrogen bonding, which increases the strength of the intermolecular forces between the Alcohol molecules, and hence raises their boiling points. Since Ethers do not have a Hydrogen bonded to their Oxygen atom, they cannot hydrogen bond to one another.

❏❏ **Unlike hydrocarbons, Ethers, especially smaller ones, are partially soluble in water. Why is this?**

Though Ether molecules are not capable of hydrogen bonding to one another, they do contain an electronegative Oxygen that possesses a partial negative charge. This negative charge is capable of interacting with the O-H dipoles of water molecules.

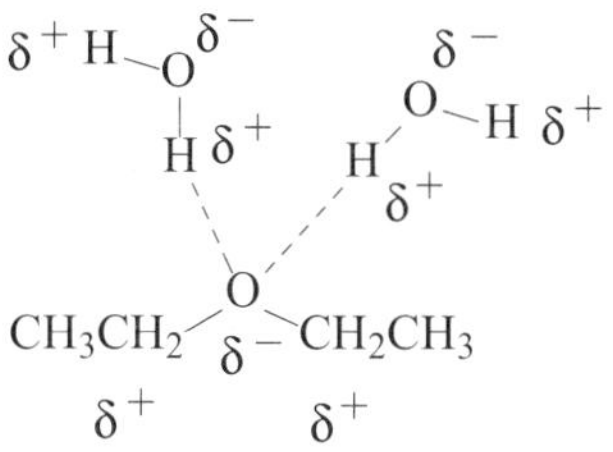

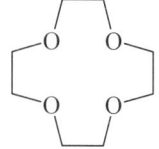

Diethyl Ether: solubility in water 6.1g / 100 g

❏❏ **Arrange the following molecules in order of their solubility in water.**

a) $CH_3OCH_2CH_2OCH_3$ b) $CH_3CH_2CH_2CH_3$

c) $CH_3CH_2OCH_2CH_3$ d) $CH_3CH_2CH_2CH_2OCH_3$

From lowest to highest solubility: b, c, d, a.

❏❏ **Ethylene Glycol Dimethyl Ether is infinitely soluble in water. Can you offer an explanation for this?**

Ethylene Glycol Dimethyl Ether has two Ether functionalities, which means two partial negative charges in a fairly small molecule. These partial charges both interact with the hydrogen bonds of water, giving the compound enhanced solubility.

❏❏ **What are Crown Ethers?**

They are cyclic polyethers formed from Ethylene Glycol or substituted Ethylene Glycol units.

❏❏ **Name the following Crown Ethers.**

a) 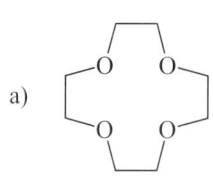   b) 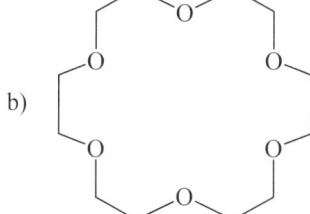   c)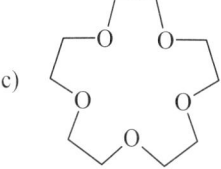

(a) 12-Crown-4, (b) 18-Crown-6, (c) 15-Crown-5

❒❒ **What is the primary use of Crown Ethers?**

Crown Ethers are used to complex alkali metal ions and pull them into solution. They are particularly useful for making inorganic salts soluble in nonpolar organic solvents such as Methylene Chloride. When the positive ions are pulled into solution, their negative counter-ions come along. In many cases this can make the anions very strong nucleophiles, because they are only weakly solvated in the organic solvents.

❒❒ **How do Crown Ethers complex alkali metal ions and bring them into solution?**

The central cavity of a Crown Ether is lined with Oxygens with unshared pairs of electrons and partial negative charges. It is therefore a very polar region and highly attractive to positively charged ions. The outer region of a Crown Ether is primarily hydrocarbon-like, and therefore highly soluble in nonpolar organic solvents. The alkali metal ions are just the right size to fit into the cavity of a Crown Ether. The Ether 18-Crown-6 is exactly suited for a Potassium ion, while 12-Crown-4 is a good fit for a Lithium ion.

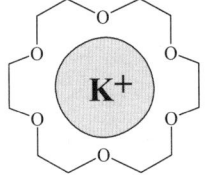

❒❒ **What is the most common method of synthesizing Ethers?**

The Williamson Ether synthesis, in which an alkoxide is reacted with an Alkyl Halide or other molecule with a good leaving group.

❒❒ **What is the mechanism of this reaction?**

The reaction proceeds by an S_N2 mechanism in which the alkoxide acts as a nucleophile and displaces the leaving group from behind.

❒❒ **For each Ether that can be made by the Williamson method, there are usually two possible approaches to the synthesis. What are they and what determines which approach is better?**

In the Williamson Ether synthesis, one side of the Ether starts off as an alkoxide nucleophile and the other side starts out as a substrate for S_N2. Theoretically, each side can play either role. From a practical standpoint, the least crowded side of the Ether is chosen to play the part of the substrate, since crowded substrates make S_N2 attack difficult.

❏❏ **How could you best synthesize the following molecules using a Williamson Ether synthesis?**

a) H₃C—CH(CH₃)—O—CH₃

b) H₃C—C(CH₃)(CH₃)—O—CH₂CH₂CH₃

c) H₃C—CH(CH₃)—O—CH₂—C(CH₃)(CH₃)—CH₃

d) CH₃CH₂—O—CH(CH₃)CH₂CH₃

a) H₃C—CH(CH₃)—O⁻Na⁺ + CH₃I

b) H₃C—C(CH₃)(CH₃)—O⁻Na⁺ + CH₃CH₂CH₂—I

c) H₃C—C(CH₃)(CH₃)—CH₂—O⁻Na⁺ + H₃C—CH(CH₃)—I

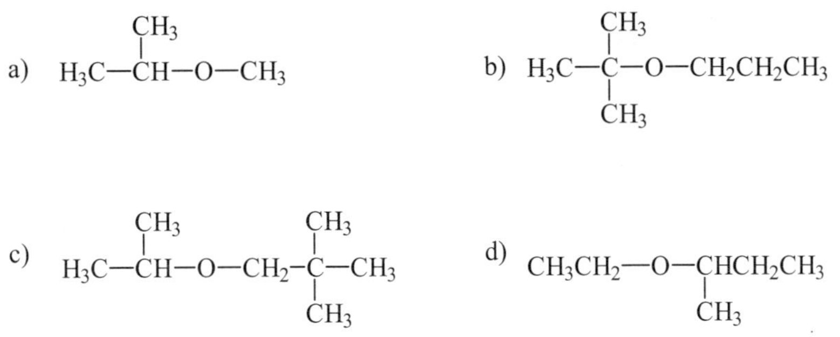

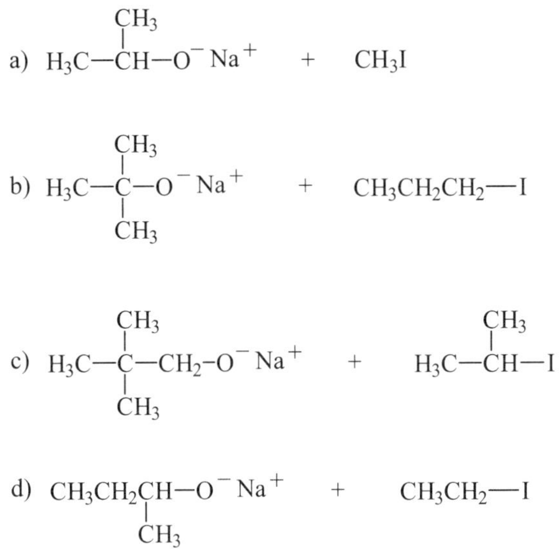

Though the other approach uses a primary substrate (Neopentyl Iodide) rather than the secondary one shown here, β-branching impedes the S_N2 mechanism.

d) CH₃CH₂CH(CH₃)—O⁻Na⁺ + CH₃CH₂—I

❏❏ **In the Williamson Ether synthesis, primary Alkyl Halides make good reactants, secondary halides give significant side products, and tertiary halides give no Ethers at all. How can this be explained?**

The principle mechanism for the Williamson Ether synthesis is S_N2. With strong nucleophiles/bases such as alkoxides, primary Alkyl Halides undergo rapid substitution, secondary Alkyl Halides react more slowly and are more likely to undergo E2 side reactions, and tertiary Alkyl Halides cannot undergo S_N2 substitution at all, and therefore give mostly elimination products.

❏❏ **In some cases, Ethers can be synthesized by adding an Alcohol to an Alkene under acidic conditions. What are the limitations on this approach?**

The Alkenes must be capable of forming stable carbocations upon addition of a proton, and the Alcohols must be primary.

❏❏ **What is the most valued chemical property of Ethers?**

They are remarkably inert and stable under a wide variety of conditions. This makes them very suitable for use as organic solvents.

❏❏ **Acid catalyzed dehydration of Ethanol has been used commercially for the production of Diethyl Ether and some other symmetric Ethers.**

$$2\ CH_3CH_2OH \xrightarrow[140\ ^\circ C]{H_2SO_4} CH_3CH_2OCH_2CH_3\ +\ H_2O$$

❏❏ **Why is this approach useless for the synthesis of unsymmetrical Ethers?**

Unsymmetrical Ethers would require the use of two different Alcohols. This would give a mixture of products.

❏❏ **What happens when secondary and tertiary Alcohols are used in an acid catalyzed synthesis of Ethers?**

The yields of Ether are lower when secondary Alcohols are used, with elimination being a significant side reaction. With tertiary Alcohols acid-catalyzed dehydration is the primary reaction.

❏❏ **What is the mechanism for the acid catalyzed addition of Alcohols to Alkenes?**

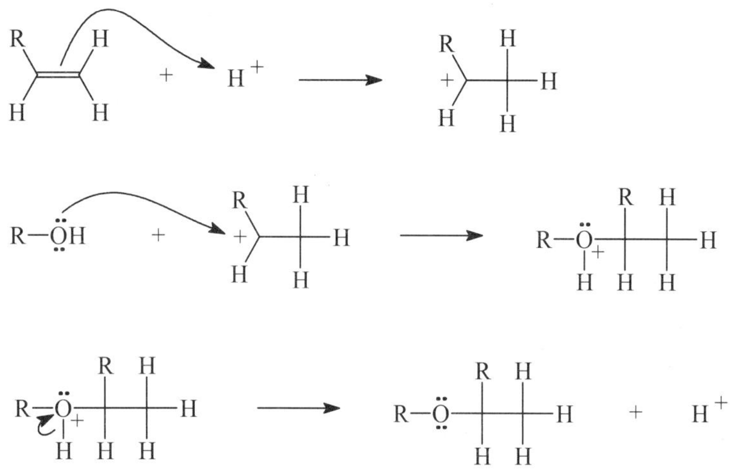

❏❏ **Under what conditions will Ethers react with Hydrogen Halides?**

In the presence of concentrated Hydrogen Halides, preferably HI or HBr.

❏❏ **What type of mechanism is followed in the cleavage of Ethers with Hydrogen Halides?**

An S_N2 mechanism, with substitution preceded by protonation of the Oxygen to make it a better leaving group.

❏❏ **What is always the first step of the cleavage reaction? Why?**

The first step is always the protonation of the Ether Oxygen to make it a better leaving group.

❏❏ **What is solvomercuration-demercuration?**

Water can be added to Alkenes to produce Alcohols by treating them first with Mercury (II) Acetate in THF/water, and then with Sodium Borohydride. This process is known as oxymercuration-demercuration. If an Alcohol is used as the solvent in the first step, instead THF/water, the product is an Ether resulting from formal addition of the Alcohol to the Alkene. This two step process is known as solvomercuration-demercuration.

❏❏ What would be the expected primary organic product(s) in each of the following reactions?

a)  CH₃—C(CH₃)(CH₃)—I + CH₃O⁻ ⟶

b) cyclopentene $\xrightarrow{\text{1) Hg(oAc)}_2, \text{CH}_3\text{OH} \quad \text{2) NaBH}_4}$

c) CH₃CH₂OH $\xrightarrow{\text{H}_2\text{SO}_4, \Delta}$

d) 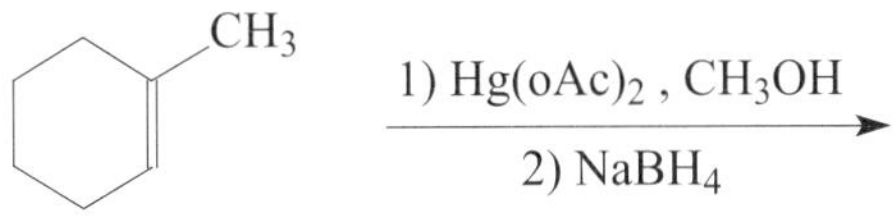 1-methylcyclohexene $\xrightarrow{\text{1) Hg(oAc)}_2, \text{CH}_3\text{OH} \quad \text{2) NaBH}_4}$

a) CH₃—C(=CH₂)—CH₃   b) cyclopentyl—OCH₃   c) CH₃CH₂—O—CH₂CH₃   d) 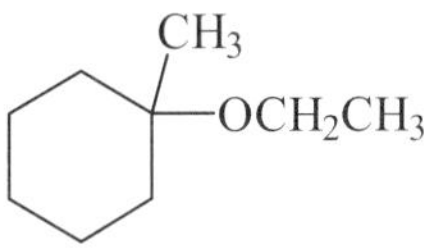 1-methyl-1-ethoxycyclohexane

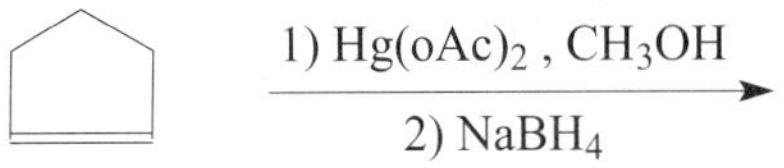

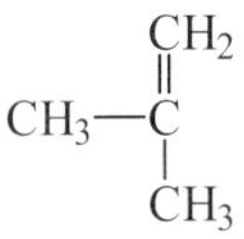

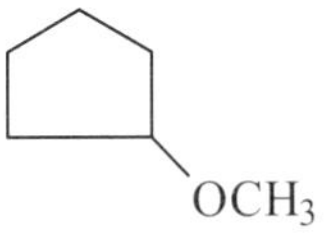

❏❏ Show the mechanism for the reaction between HI and Diethyl Ether.

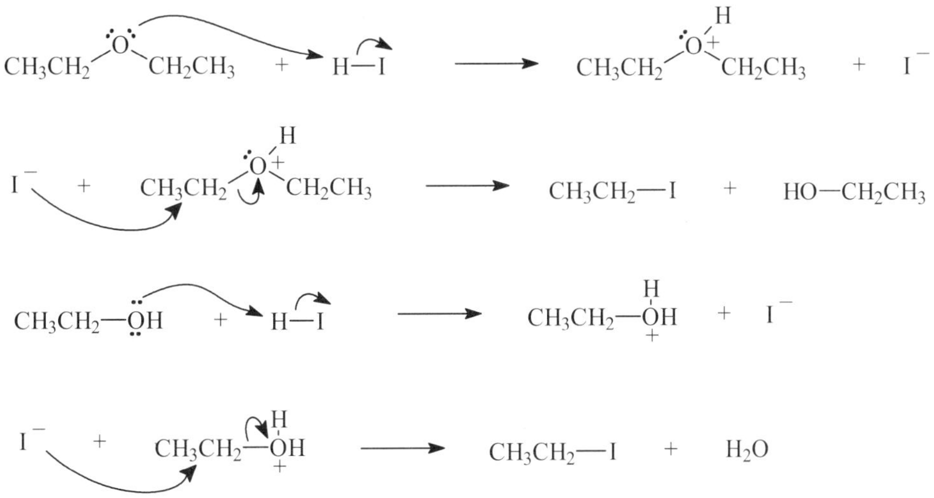

CH₃CH₂—Ö—CH₂CH₃ + H—I ⟶ CH₃CH₂—Ö⁺(H)—CH₂CH₃ + I⁻

I⁻ + CH₃CH₂—Ö⁺(H)—CH₂CH₃ ⟶ CH₃CH₂—I + HO—CH₂CH₃

CH₃CH₂—ÖH + H—I ⟶ CH₃CH₂—Ö⁺H₂ + I⁻

I⁻ + CH₃CH₂—Ö⁺H₂ ⟶ CH₃CH₂—I + H₂O

❏❏ **Why is unwise to keep around old bottles of Ether?**

Old bottles of Ether may contain Hydroperoxides, which are explosive.

❏❏ **What is autoxidation?**

Oxidation by Oxygen in the air. The process occurs by a radical mechanism.

❏❏ **What type of Ether is most at risk for formation of Hydroperoxides?**

Ethers with a tertiary Hydrogen on the Carbon adjacent to the Oxygen.

❏❏ **How are Hydroperoxides in Ether detected?**

By shaking the Ether with an acidified solution of 10% aqueous Potassium Iodide. Hydroperoxides oxidize the Iodide (I$^-$) to Iodine (I$_2$), which gives the solution a yellow color. If Starch is added then the deep bluish-purple color of the Starch-Iodine complex appears.

❏❏ **How can Hydroperoxides be removed from Ethers?**

By treatment with a reducing agent such as Iron (II) Sulfate in dilute aqueous Sulfuric Acid.

❏❏ **What is a protecting group?**

During a synthesis, it is sometimes necessary to protect a functional group from reaction conditions necessary to make changes in other parts of the molecule. This can be done by making temporary changes in the nature of the functional group so as to render it inert to the reagents to be used. Such a change in functionality is known as a protecting group. To be useful, protecting groups must be easy to introduce, completely stable under the specific reaction conditions that will be used, and easily removed when the synthesis step is completed.

❏❏ **Why are Ethers often used as protecting groups?**

Because Ethers are remarkably stable under a wide variety of reaction conditions, including oxidative and basic conditions.

❏❏ **How would you protect an Alcohol as an Ether?**

You could easily convert the Alcohol to an Ether through a Williamson Ether synthesis. Reactions involving other functional groups in the molecule could then be carried out without the Alcohol reacting.

❏❏ **How is an Ether protecting group removed?**

By treatment with aqueous acid.

❏❏ **What is a THP protecting group?**

THP stands for Tetrahydropyranyl. THP protecting groups are made by reacting an Alcohol to be protected with Dihydropyran in the presence of anhydrous acid catalysts. They are removed with aqueous acids.

❏❏ **Why are Epoxides so much more reactive than other Ethers?**

Though their functionalities appear similar, Epoxides have a three membered ring. This three membered ring places a considerable strain on the molecule, which can be relieved by any ring opening reaction.

❏❏ **Give the structures of the following compounds.**

a) Oxirane **b) 1,2-Dimethylcyclopentene oxide**

c) (R)-2,2,3-Trimethyloxirane **d) 5,6-Epoxyhexane**

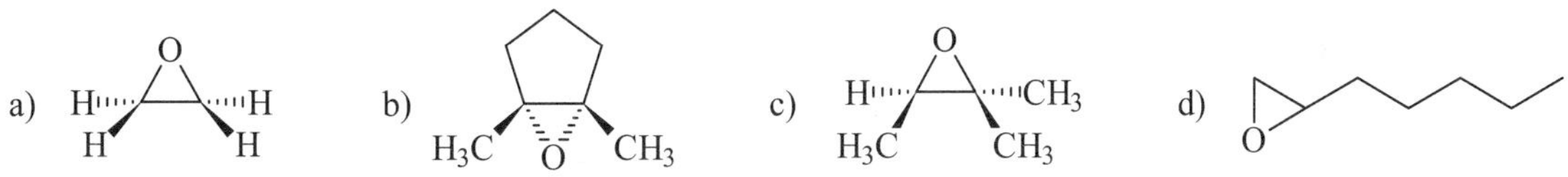

❏❏ **How are Epoxides usually synthesized?**

By treating an Alkene with a Peroxy-acid.

❏❏ **What is the mechanism of this epoxidation reaction?**

The mechanism is a one step process:

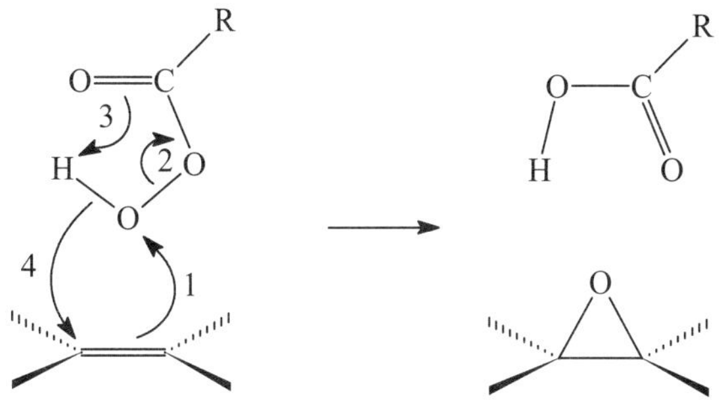

❏❏ **Is this reaction Stereospecific?**

Yes, very. The Oxygen is added Syn to one side of the Alkene. Any substituents that were on the same side of the Alkene before the reaction find themselves on the same side in the product.

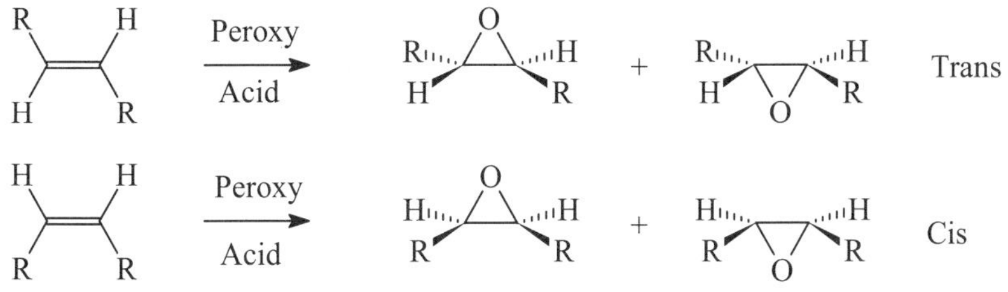

❏❏ **What other common methods can be used to create Epoxides?**

Epoxides can be made by treating halohydrins with base. This results in an intramolecular S_N2 reaction.

❏❏ **What would be the expected primary organic product(s) in the following reactions? Be sure to indicate stereochemistry where appropriate.**

a) 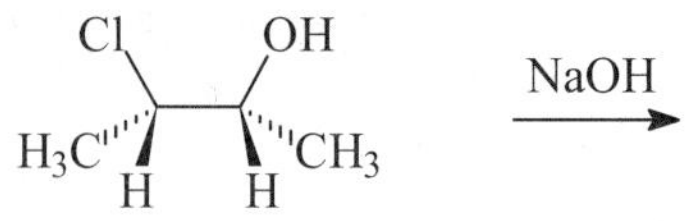 [structure: (2R,3S) or meso-like 2-chloro-3-hydroxybutane with Cl and OH on adjacent carbons, H₃C and CH₃ wedged, H's dashed] → NaOH

b) 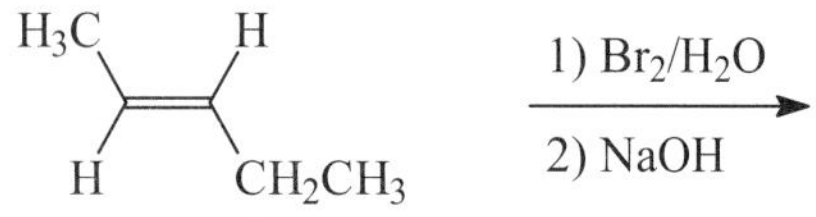 [structure: (Z)- or cis-2-pentene with H₃C and H on one carbon, H and CH₂CH₃ on the other] → 1) Br₂/H₂O 2) NaOH

c) 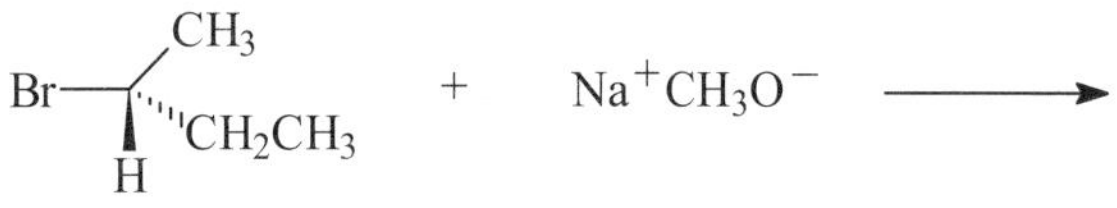 Br—C(CH₃)(CH₂CH₃)(H) + Na⁺CH₃O⁻ →

a) 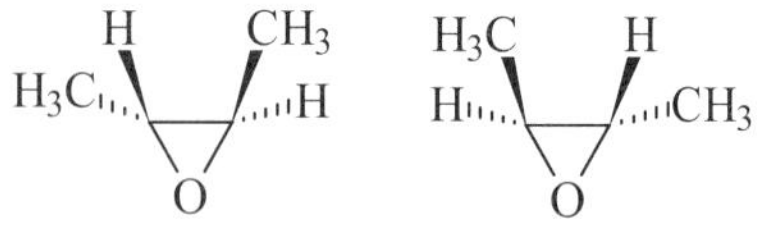 [two epoxide structures: H₃C/H and CH₃/H on one; H₃C/H and H/CH₃ on the other]
   Racemic Mixture

b) 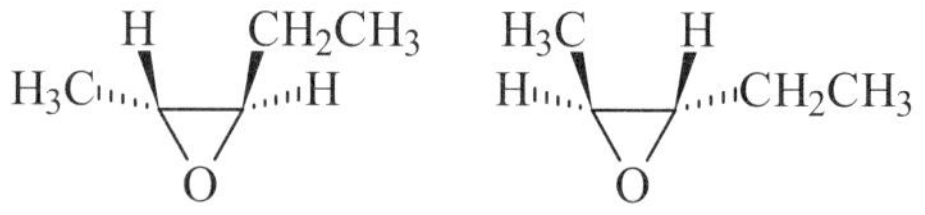 [two epoxide structures with H/CH₂CH₃ and CH₃/H; and H₃C/H and H/CH₂CH₃]
   Racemic Mixture

c) 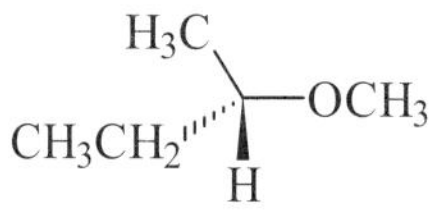 CH₃CH₂—C(H)(CH₃)—OCH₃

❏❏ **When nucleophiles react with Epoxides, what stereochemistry and regiochemistry is observed?**

The stereo and regiochemistry depend on the reaction conditions. Under neutral or basic conditions, the nucleophile will attack at the least hindered Carbon, and attack from the face opposite the Epoxide Oxygen. The configuration of the Carbon that is attacked is inverted as in any S$_N$2 reaction. Under acid catalyzed conditions, the Epoxide Oxygen is protonated, and when the nucleophile attacks, considerable carbocation character is placed on the Carbon being attacked. The nucleophile therefore attacks at the most substituted Carbon. The attack still occurs from the face opposite that of the Epoxide Oxygen and stereochemistry still proceeds with inversion, however.

Nu → [epoxide with H, H, H, R] → [product: H, OH, Nu, H, R] Neutral or Basic Conditions

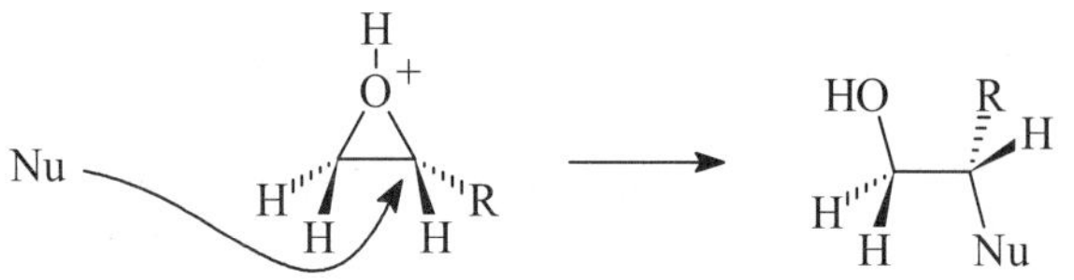

Nu → [protonated epoxide with H, H, H, R] → [product: HO, R, H, Nu, H] Acidic Conditions

❏❏ **When Sulfuric Acid is used as a catalyst, Epoxides are cleaved to give 1,2-Diols. What is the mechanism of this reaction?**

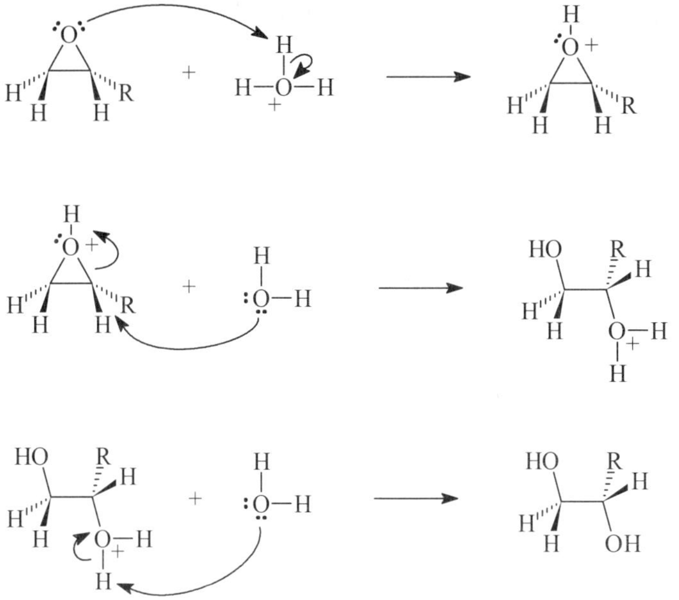

Nucleophilic attack occurs at the most substituted carbon because of the partial cationic character in the transition state.

❏❏ **What is the regioselectivity of the ring opening of Epoxides with strong nucleophiles?**

Strong nucleophiles are usually used under neutral or basic conditions and under those conditions attack occurs at the least substituted Carbon of the Epoxide

❏❏ **Epoxides can be chemically reduced to give Alcohols. What reagent is used to perform this transformation?**

Lithium Aluminum Hydride, $LiAlH_4$.

❏❏ **What is the regioselectivity of this reaction?**

The hydride is transferred to the least crowded Carbon.

❏❏ **What is the Sharpless epoxidation reaction?**

It is an enantioselective epoxidation reaction that can be used to produce a single Enantiomeric Epoxide from an Allyl Alcohol. The reagent used in the reaction is a mixture of *tert*-Butylhydroperoxide, Titanium (IV) Isopropoxide, and Diethyl (2R, 3R)-tartrate.

❏❏ **Upon treatment with hydriodic acid at elevated temperatures, an unknown Ether yields 1,6-Diiodohexane as the sole product. What is the probable structure of the Ether?**

The product is most likely a cyclic Ether, one with six Carbons and one Oxygen in the ring.

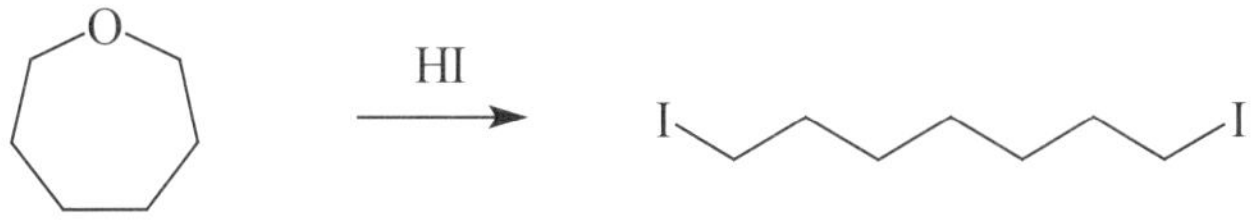

THIOL, SULFIDE, AND DISULFIDE PEARLS

Stir a cesspit and a foul stench arises;
stir perfume and a delightful fragrance ascends.
But the movement is identical.
St. Augustine

❑❑ **What is a Thiol?**

A Thiol is a compound containing an -SH group bonded to an sp^3 hybridized Carbon.

❑❑ **What older name is also used for this class of compounds?**

They are also known as Mercaptans.

❑❑ **Name the following compounds using IUPAC standard nomenclature**

a) 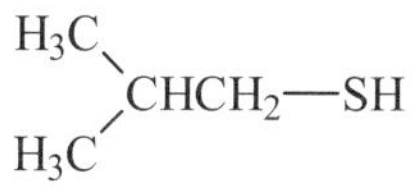 (CH$_3$)$_2$CHCH$_2$—SH

b) 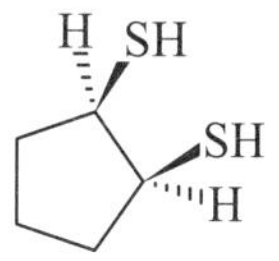 cis-cyclopentane with two SH groups

c) 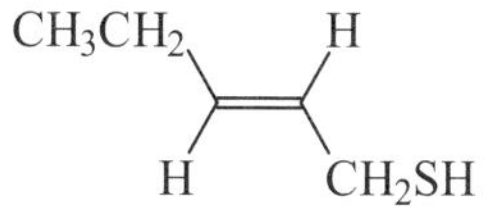 (E)-CH$_3$CH$_2$CH=CHCH$_2$SH

d)  CH$_3$CH(SH)CH$_2$CH$_2$—OH

(a) 2-Methyl-1-propanethiol, (b) *cis*-1,2-Cyclopentanedithiol, (c) (E)-2-Pentene-1-thiol, (d) 3-Mercapto-1-butanol

❑❑ **Give the common names of the following compounds.**

a) 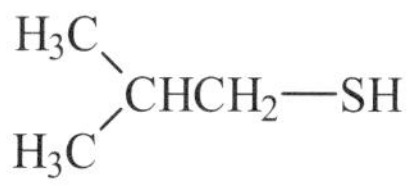 (CH$_3$)$_2$CHCH$_2$—SH

b) HS—CH$_2$CH$_2$—SH

c) 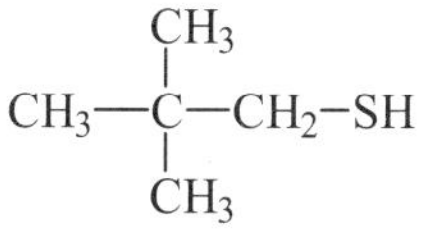 (CH$_3$)$_3$C—CH$_2$—SH

d) 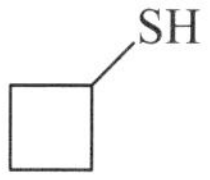 cyclobutyl-SH

(a) Isobutyl Mercaptan, (b) Ethylene Dimercaptan, (c) Neopentyl Mercaptan, (d) Cyclobutyl Mercaptan

❏❏ **An -OH is known as an Hydroxyl group. What is an -SH group called?**

A Sulfhydryl group.

❏❏ **What is a sulfide?**

A Sulfide is a compound with a Sulfur atom bonded to two Carbon atoms. It is the Sulfur analog of an Ether.

❏❏ **What is the most common method of preparing symmetrical Sulfides?**

Symmetrical sulfides are usually prepared by treating Sodium Sulfide (Na_2S) with two equivalents of Alkyl Halide. The Sulfide ion (S^{2-}) acts as a nucleophile.

$$Na^+_2 S^{2-} + 2\ R-X \longrightarrow R-S-R + 2\ NaX$$

❏❏ **How are unsymmetric Sulfides prepared?**

By deprotonating a Thiol with a strong base such as Sodium Hydroxide, and then reacting the resulting salt with an Alkyl Halide.

$$Na^+\ R-S^- + R'-X \longrightarrow R-S-R' + NaX$$

❏❏ **What other name is there for Sulfides?**

They are also called Thioethers.

❏❏ **Give the IUPAC names of the following Sulfides.**

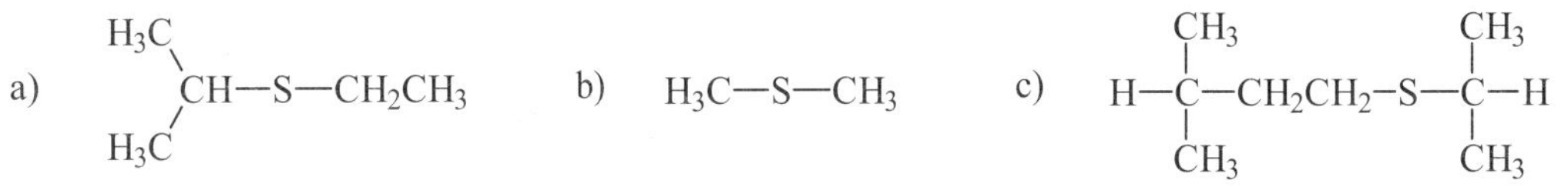

(a) Ethyl Isopropyl Sulfide, (b) Dimethyl Sulfide, (c) Isopentyl Isopropyl Sulfide

❏❏ **How could you best prepare the following Sulfides?**

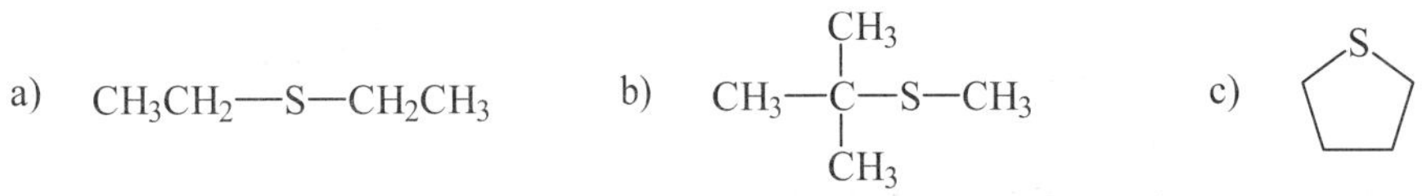

(a) React Disodium Sulfide with two equivalents of Iodoethane, (b) treat *tert*-Butyl Mercaptan with Sodium Hydroxide to deprotonate it, then react the salt with Methyl Iodide, (c) react Disodium Sulfide with one equivalent of 1,4-Diiodobutane.

❏❏ **What is the most characteristic property of Thiols?**

Their incredibly unpleasant odor.

❏❏ **Thiols show much lower boiling points than those of similar Alcohols. Why is this?**

The Sulfur atom is not very electronegative, so the S-H bond is not particularly polar. As a result, Thiols are not capable of hydrogen bonding the way that Alcohols do, and they end up having lower boiling points.

☐☐ **How would you expect the boiling points of a Thiol and an Isomeric Sulfide to compare?**

Since the Thiol and the Sulfide have the same size and the same atoms, and the Thiol gains no additional intermolecular attractions by having an S-H bond, they should have very similar boiling points. In fact, the boiling points of Ethanethiol and Dimethyl Sulfide differ by only two degrees.

CH_3—S—CH_3 CH_3CH_2—SH

bp = 37 ºC   bp = 35 ºC

☐☐ **How do the solubilities in water of Thiols and Alcohols compare?**

Unlike Alcohols, Thiols have very low solubilities in water - even small Thiols. This is again because Thiols are incapable of hydrogen bonding.

☐☐ **What is the most common way of preparing Thiols?**

Thiols are usually prepared by bubbling H_2S through a solution of Sodium Hydroxide in water or Ethanol. This forms Hydrosulfide ion, HS^-, which is then reacted with an Alkyl Halide to produce the Thiol. An excess of Hydrosulfide ion is typically used.

☐☐ **What are the limitations of this approach?**

The reaction works best with primary Alkyl Halides, which of course results in only primary Thiols. When secondary and tertiary Alkyl Halides are used, β-elimination competes with substitution, giving significant elimination products with secondary halides, and solely elimination products with tertiary halides.

☐☐ **Why must an excess of Hydrosulfide Anion be used?**

Because the Thiol product will also react with the Alkyl Halide, to give an undesired symmetrical Sulfide.

☐☐ **How do the acidities of Thiols and Alcohols compare?**

Thiols are much more acidic than Alcohols, and may even be deprotonated in aqueous solutions.

☐☐ **What is the typical pK_a of a Thiol?**

A typical Alkane Thiol has a pKa of about 10.

☐☐ **Though Thiols are largely insoluble in water, there is a very easy way to make them dissolve. What is it?**

Add a base to deprotonate them. Thiols themselves may be insoluble, but their salts are not.

☐☐ **When Alcohols are oxidized, Aldehydes, Ketones, or Carboxylic Acids are the result. What happens when Thiols are oxidized?**

In Thiols, no C=S bond is formed through oxidation. The sulfur is actually oxidized in preference to the Carbon. This results in sulfenic, sulfinic, and sulfonic acids

R—SH $\xrightarrow{Ox.}$ R—S—OH   R—S(=O)—OH   R—S(=O)(=O)—OH

    Sulfenic   Sulfinic   Sulfonic
    Acid   Acid   Acid

❏❏ **How would you expect OH⁻ and SH⁻ to compare as bases?**

Since H_2S is a stronger acid than water (because it can be deprotonated in water), the conjugate base SH⁻ must be a weaker base than hydroxide.

❏❏ **What is a Disulfide?**

A disulfide is a compound containing an S-S group.

❏❏ **How easy is it to oxidize a Thiol?**

Thiols may be oxidized even by atmospheric Oxygen.

❏❏ **What are the two most common situations in which to encounter Thiols?**

People often encounter Thiols when using a gas stove (a Thiol is added to natural gas to give it its characteristic smell), or when driving by dead skunks on the highway (the smell of skunks is the result of two Thiols).

❏❏ **What is Ethanethiol used for?**

Ethanethiol is the substance added to natural gas to give it a smell.

❏❏ **How sensitive are we to the smell of Thiols?**

Typical Thiols can be detected at levels as low as 0.02 parts per billion (that's 0.02 parts Thiol per billion parts air).

❏❏ **As Thiols increase in the number of Carbons, their odor decreases. Why is this?**

Because as the number of Carbons increases, the volatility of the compound drops, and so does the percentage of sulfur.

❏❏ **How are Thiols usually converted to Disulfides?**

By using a mild oxidizing agent such as Br_2, I_2, or $K_3Fe(CN)_6$.

❏❏ **When Thiols are oxidized to Disulfides, how can they be converted back?**

With a reducing agent such as Zn in acid or Li in NH_3.

❏❏ **Why are Thiol-Disulfide conversions important in biochemistry?**

Disulfide bridges cross-link between protein chains and help to stabilize the three dimensional structure of proteins.

❏❏ **What do Thiols and hair have to do with one another?**

Hair has many S-H groups. When a person gets a perm, some of these groups are oxidized to form disulfide links, locking the hair into a new, wavier shape.

❏❏ **What is a Sulfoxide?**

$$R-\underset{\underset{R}{}}{\overset{\overset{O}{\|}}{S}}-R$$

a Sulfoxide

❏❏ **What is a Sulfone?**

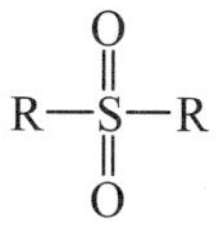

$$R-\underset{\underset{O}{\|}}{\overset{\overset{O}{\|}}{S}}-R$$

a Sulfone

❏❏ **Under what conditions are Sulfides oxidized to Sulfoxides?**

Converting a Sulfide to a Sulfoxide requires 30% Hydrogen Peroxide in the presence of acid catalyst at 25 °C.

❏❏ **Under what conditions are Sulfones the result?**

If the reaction is run at 100 °C, Sulfones are obtained.

❏❏ **What Amino Acid is responsible for cross-linking in proteins?**

Cysteine.

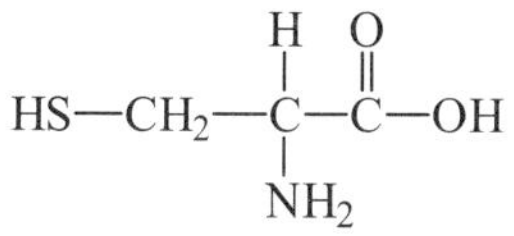

$$HS-CH_2-\underset{NH_2}{\overset{H}{\underset{|}{C}}}-\overset{O}{\overset{\|}{C}}-OH$$

❏❏ **What products are obtained when Thiols are reacted with Aldehydes and Ketones?**

Thioacetals.

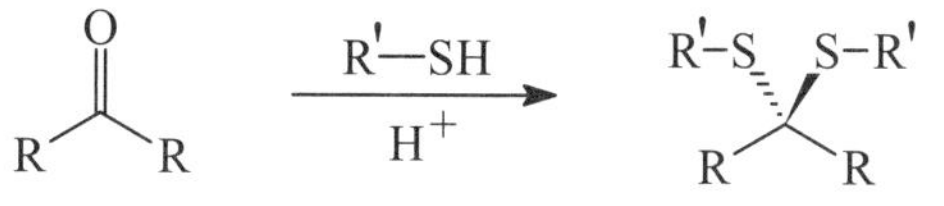

R(C=O)R →[R'-SH, H⁺] R'S-C(R)(R)-SR'   a Thioacetal

❏❏ **Why is this reaction product useful?**

Thioacetals are useful because they can be desulfurized with Raney Nickel to give Alkanes. This provides a simple way to reduce Ketones and Aldehydes to Alkanes under neutral conditions.

R'S-C(R)(R)-SR' →[Raney Ni] H-C(R)(R)-H

❏❏ **Ethers are fairly unreactive except under acidic conditions. Sulfides, however, make good nucleophiles. Why is this?**

The valence electrons of Sulfur are less tightly held than those of Oxygen because they are in a higher shell. This makes Sulfur in Sulfides much more polarizable, and therefore more nucleophilic, than the Oxygen in Ethers.

❏❏ **When Sulfides are reacted with Alkyl Halides, what type of product is obtained?**

A Trialkylsulfonium salt.

❏ ❏ **The compound Methyl-p-tolyl Sulfoxide has a chiral Sulfur, even though only three atoms are bonded to it. How is this possible?**

Though only three atoms are bonded to the Sulfur, it is chiral because it also has a lone pair that occupies one corner of a tetrahedron. Unlike the lone pair on a Nitrogen, which constantly undergoes pyramidal inversion, the lone pair on a sulfoxide Sulfur is fixed in place.

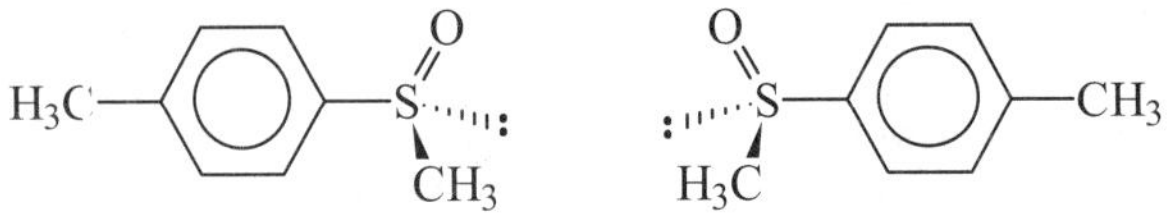

Methyl-p-tolyl Sulfoxide

❏ ❏ **What is DMSO?**

DMSO stands for Dimethyl Sulfoxide. It is commonly used as a polar aprotic solvent.

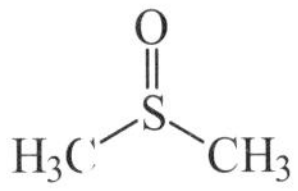

DMSO

❏ ❏ **Grignards react with Disulfides to give Sulfides as products. But when Grignards are reacted with Thiols, the result is decomposition of the Grignard. Why is this?**

Thiols have acidic Hydrogens. Reaction of any acid with a Grignard reagent results in its decomposition.

❏ ❏ **Which would be a better nucleophile, an alkoxide anion (RO⁻) or a thiolate anion (RS⁻)?**

A Thiolate Anion. Thiolate Anions are among the best nucleophiles known.

AROMATIC COMPOUND PEARLS

My mental eye, rendered more acute by repeated visions of this kind, could now distinguish larger structures of manifold conformations; long rows, sometimes more closely fitted together; all twisting and turning in snake-like motion. But look! What was that? One of the snakes had seized hold of its own tail, and the form whirled mockingly before my eyes. As if by a flash of lightning I woke.... Let us learn to dream gentlemen, and then perhaps we shall learn the truth.
Friedrich August Kekulé (1890)

❏❏ **Who first isolated Benzene?**

Michael Faraday.

❏❏ **What is the index of Hydrogen deficiency of Benzene?**

Four.

❏❏ **What shape is Benzene?**

It is perfectly hexagonal in shape, and planar.

❏❏ **How is the structure of Benzene typically depicted?**

There are two common ways of representing the structure of Benzene. One way is as a six membered ring with alternating single and double bonds. The other is as a six membered ring with a circle in the middle, which stresses the idea that all of the C-C bonds are equivalent and the Benzene structure can be best discribed as a resonance hybrid.

❏❏ **When the chemistry of Benzene was first investigated, what was it that was found to be remarkable?**

Given its high degree of unsaturation, Benzene is remarkably stable. It does not undergo any of the oxidation, reduction, or addition reactions that are characteristic of other unsaturated compounds such as Alkenes and Alkynes.

❏❏ **What is the most characteristic reaction of Benzene?**

Substitution, in which a group replaces one of the Hydrogens of Benzene.

❏❏ **Who first proposed the structure of Benzene?**

Friedrich August Kekulé in 1865 proposed that Benzene was a six-membered ring with one Hydrogen attached to each Carbon. In 1872 he suggested that the ring contained three double bonds that shifted position so quickly that the two forms could not be separated.

❏❏ **By what name is the Benzene structure with the alternating double bonds known?**

The two possible structures are known as Kekulé structures.

❑❑ **Why did Kekulé propose the rapid shifting of the double bonds?**

When Benzene is disubstituted, as with Benzene during bromination, only three isomers are isolated, with 1,2 substitution, 1,3 substitution, and 1,4 substitution. If in fact the double bonds were fixed in place, a total of four isomers would be expected, because 1,2 substitution would be different from 1,6 substitution. Rapid shifting of the double bonds removes this problem.

Three disubstituted isomers observed

No double bond between substituents ≠ Double bond between substituents

❑❑ **What are the Carbon-Carbon bond lengths in Benzene?**

All of the Carbon-Carbon bonds are the same length: 140 pm. This is halfway between the length of a typical double bond (134 pm) and a typical single bond (146 pm).

❑❑ **Because Benzene does not react like most unsaturated compounds it is said to be stabilized. How can the extent of this stabilization be measured?**

The stabilization can be estimated from heats of hydrogenation. Cyclohexene has a heat of hydrogenation of 120 kJ/mol. The heat of hydrogenation of 1,3-Cyclohexadiene is 231 kJ/mol, a little less than 2x120. If Benzene actually behaved as though it had alternating single and double bonds (that is, as though it were 1,3,5-Cyclohexatriene), the expected heat of hydrogenation would be about 3 x 120 = 360 kJ/mol. The actual measured heat of hydrogenation of Benzene is only 208 kJ/mol. This means that Benzene is stabilized by 152 kJ/mol (360-208) over the hypothetical structure 1,3,5-Cyclohexatriene. This quantity is known as the Empirical Resonance Energy.

❑❑ **What is aromaticity?**

Aromaticity refers to the special stability of compounds that meet the criteria for being aromatic. Aromatic compounds must be cyclic, flat, have a p orbital on each atom of the ring, and have a closed loop of $4n + 2$ pi electrons in the cyclic arrangement of p orbitals.

❑❑ **What is Hückel's Rule?**

Completely conjugated planar monocyclic hydrocarbons have special stability when their π systems contain $4n+2$ electrons, where n is an integer.

❑❑ **What is an antiaromatic compound?**

A antiaromatic compound is any completely conjugated planar monocyclic compound that has $4n$ electrons (n is an integer) in its π system. Such compounds are especially unstable.

❑❑ **Cyclobutadiene and Cyclooctatetraene are both antiaromatic. Cyclobutadiene is unstable and can only be observed at temperatures close to absolute zero, while Cyclooctatetraene is stable and has properties similar to Alkenes. How is this possible?**

Cyclobutadiene is a planar compound with $4n$ (n = 1) electrons. It therefore shows the low stability expected of antiaromatic compounds. Cyclooctatetraene, however, is capable of assuming a non-planar structure due to its large ring size. It thus manages to escape the instability associated with antiaromaticity by destroying the conjugation of the double bonds. The compound is therefore stable, and the double bonds react just like isolated Alkenes.

❏❏ What is the Frost Circle (or inscribed polygon) method?

It is a graphical method for determining the relative energies of the π molecular orbitals of a fully conjugated planar monocyclic hydrocarbon. The method consists of drawing a circle and then drawing a polygon within the circle that has the same number of sides as the cyclic hydrocarbon in question. The polygon must be drawn so that one of the vertices intersects the bottom of the circle. The relative energies of the p molecular orbitals may be found by examining the points at which the vertices contact the circle.

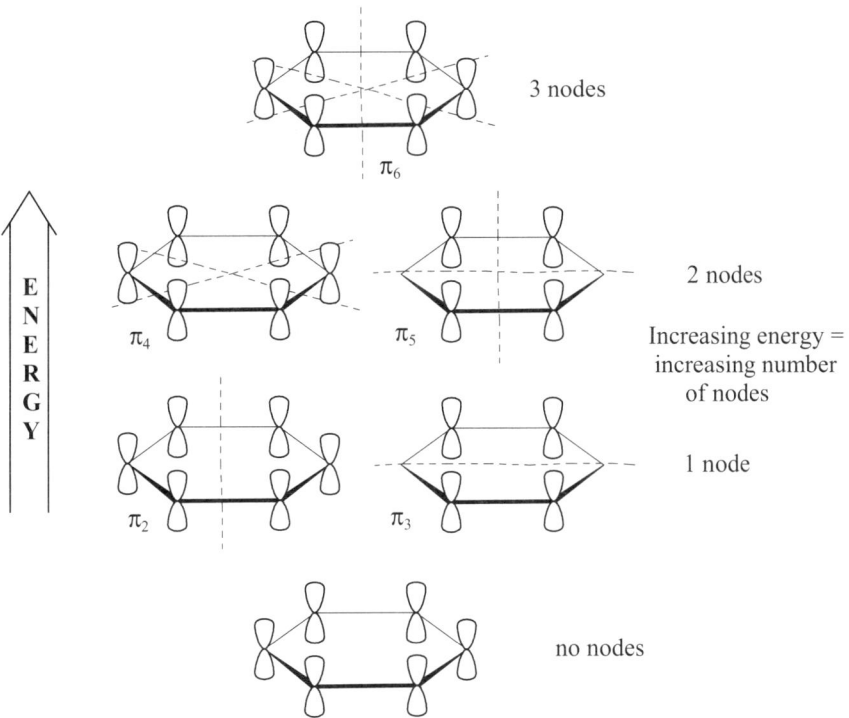

❏❏ Draw all of the π molecular orbitals of Benzene.

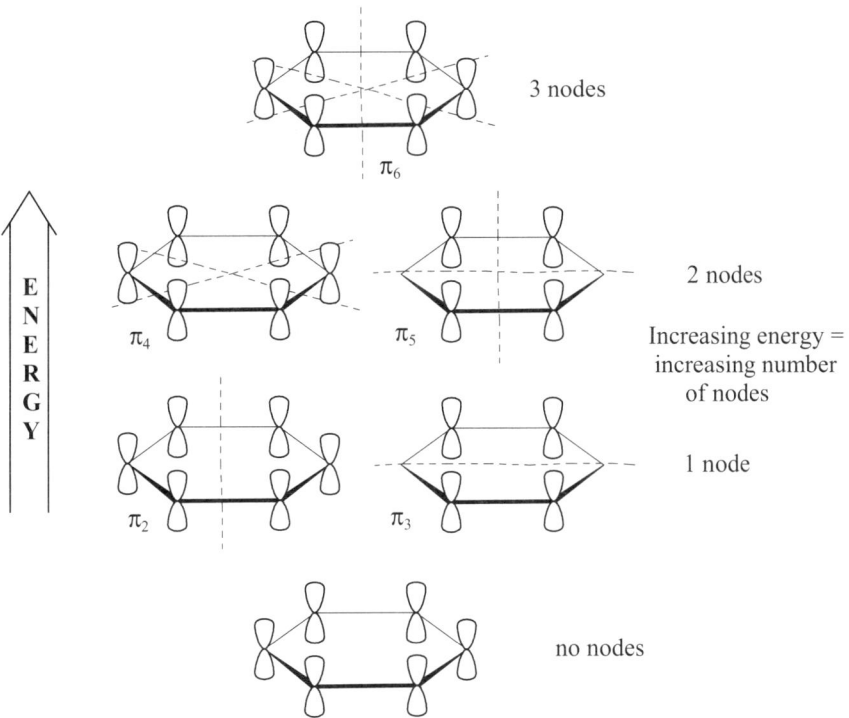

❏❏ What is an Annulene?

An Annulene is a cyclic hydrocarbon with a continuous series of alternating single and double bonds. Benzene is a member of the Annulene family.

❏❏ How are large Annulenes named?

The number of Carbons in the ring is given in square brackets followed by the word Annulene. Benzene could be named as [6] Annulene.

❏❏ **What would be the IUPAC name for each of the following molecules?**

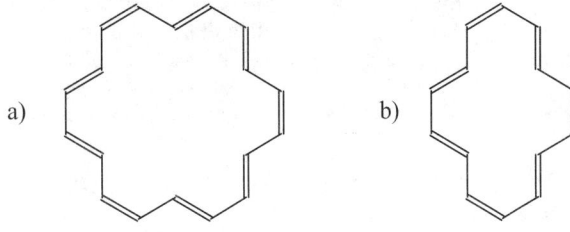

(a) [18] Annulene, (b) [14] Annulene, (c) 1,2-Dibromobenzene, (d) Naphthalene

❏❏ **What do ortho, para, and meta mean?**

Ortho, meta, and para are locators used to describe the relative positions of substituents on disubstituted Benzene rings. Para means that the substituents are in a 1,4 position, meta means that they are in a 1,3 position, and ortho means that they are in a 1,2 position.

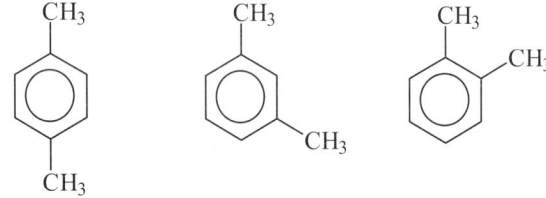

Para Xylene   Meta Xylene   Ortho Xylene

❏❏ **Would each of the following compounds be considered aromatic, antiaromatic, or non-aromatic?**

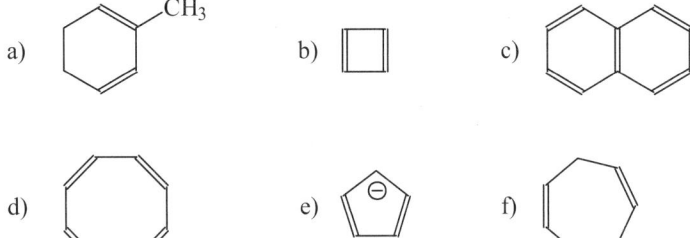

(a) Aromatic, (b) Antiaromatic, (c) Aromatic, (d) Antiaromatic, (e) Aromatic, (f) Non-aromatic

❏❏ **What would the common names be for each of the following Benzene derivatives?**

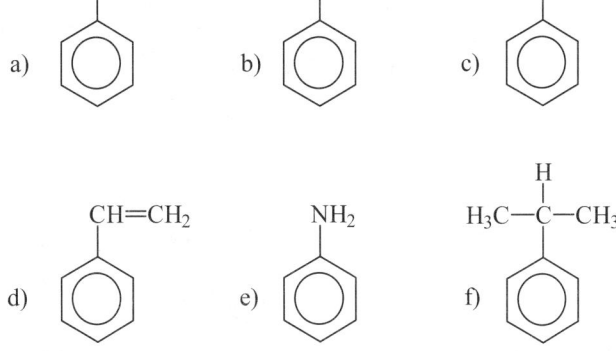

(a) Toluene, (b) Phenol, (c) Anisole, (d) Styrene, (e) Aniline, (f) Cumene

❏❏ **Give two IUPAC standard names for each of the following compounds.**

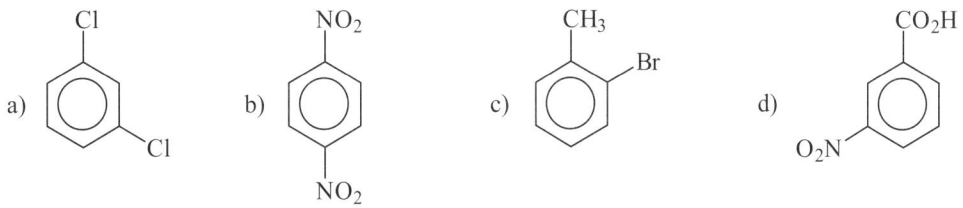

(a) 1,3-Dichlorobenzene or *m*-Dichlorobenzene, (b) 1,4-Dinitrobenzene or *p*-Dinitrobenzene, (c) 2-Bromotoluene or *o*-Bromotoluene, (d) 3-Nitrobenzoic acid or *m*-Nitrobenzoic acid

❏❏ **What is a polynuclear aromatic compound?**

They are compounds that contain two or more fused Benzene rings.

❏❏ **What would be the IUPAC name for each of the following compounds?**

(a) 2-Benzyl-1-pentanol, (b) 2,3,5-Trichlorophenol, (c) 4-Methylstyrene or *p*-Methylstyrene

❑❑ Which of the following heterocycles are considered to be aromatic?

a) 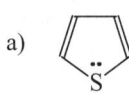 b) 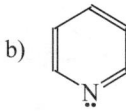 c)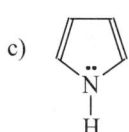

d) 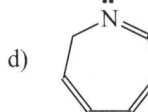 e) 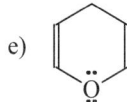 f)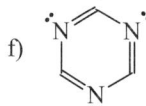

(a) Aromatic, (b) Aromatic, (c) Aromatic, (d) Non-aromatic, (e) Non-aromatic, (f) Aromatic

❑❑ Most carbocations are unstable, yet **Cyclopropenyl Tetrafluroborate**, a carbocation salt, can be isolated and stored in bottles. Why is this?

The Cyclopropenyl Carbocation is cyclic, planar, has p orbitals on each of the ring Carbons, and has two electrons in its π system (4n + 2, n = 0.) It therefore meets all of the criteria of an aromatic compound and shows the predicted stability. Any reaction that destroys the carbocation would also destroy the aromaticity.

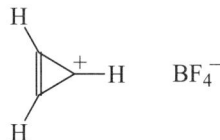

❑❑ The unusual stability of Benzene is often described in terms of simple resonance, based on the fact that Benzene has two equivalent resonance forms. Why is this an inadequate explanation?

If the number of resonance forms were the only criteria for aromaticity, then Cyclobutadiene would also be aromatic because it possesses two equivalent resonance forms. Cyclobutadiene, however, is antiaromatic

❑❑ What are the IUPAC names of the following aromatic hetereocycles?

a) 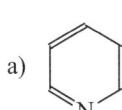 b) 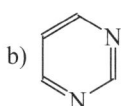 c) 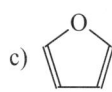 d)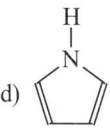

(a) Pyridine, (b) Pyrimidine, c) Furan, (d) Pyrrole

❑❑ What is a Tropylium ion?

A Tropylium ion is a Cycloheptatrienyl cation. It is a planar seven-membered ring cation with p orbitals on each Carbon of the ring and six electrons in the π system. It is therefore aromatic.

❏❏ **What would be the structure of each of the following compounds?**

a) 4-Hydroxy-3-methoxybenzaldehyde (Vanillin)
b) 1,4-Benzenediol(hydroquinone)
c) 1,3-Dimethylbenzene (*m*-Xylene)
d) Anthracene

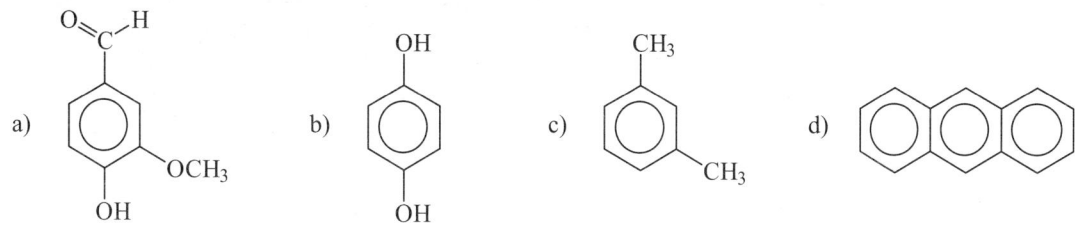

❏❏ **Alcohols are usually not very acidic (pK$_a$ of about 16), but phenol has a pK$_a$ of 9.9. Why is this?**

The anion of an aliphatic Alcohol has only one resonance form. The negative charge is therefore localized strictly on the Oxygen. A Phenolate ion has four resonance structures, spreading the charge out over the molecule and making it much more stable. Since the anion is more stable, it is easier to form and this is reflected in the pK$_a$ value of the conjugate acid.

❏❏ **What would be the order of acidities for the following compounds?**

From strongest to weakest acid: a, d, e, c, b.

❏❏ **What is Electrophilic Aromatic Substitution?**

Electrophilic Aromatic Substitution is a reaction in which an Electrophile is substituted for a Hydrogen on an aromatic ring.

❏❏ **What is an Ortho-Para Director?**

An Ortho-Para Director is a substituent on a Phenyl ring that directs all subsequent substitution to either the ortho or para positions.

❏❏ **What is the typical mechanism for Electrophilic Aromatic Substitution?**

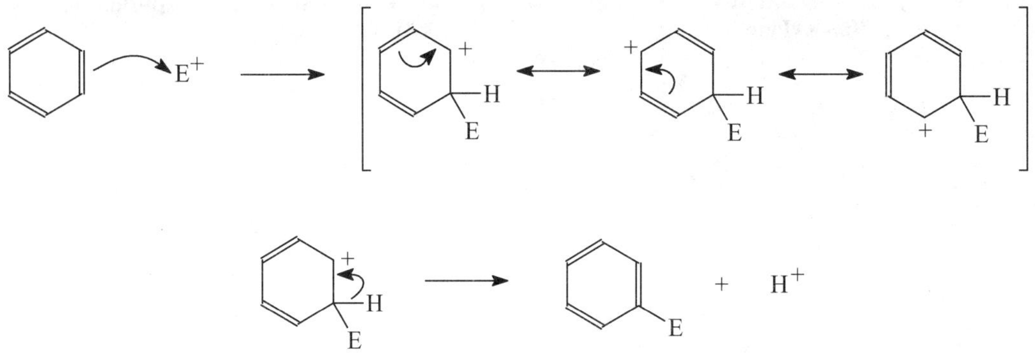

❏❏ **What would you expect to be the primary organic product(s) from each of the following reactions?**

a) 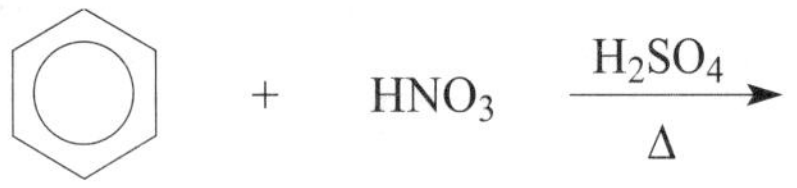 benzene + HNO_3 $\xrightarrow{H_2SO_4, \Delta}$

b) 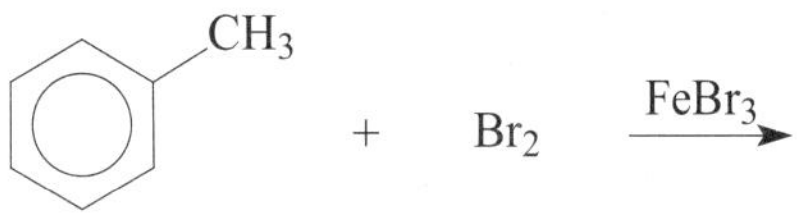 toluene + Br_2 $\xrightarrow{FeBr_3}$

c) 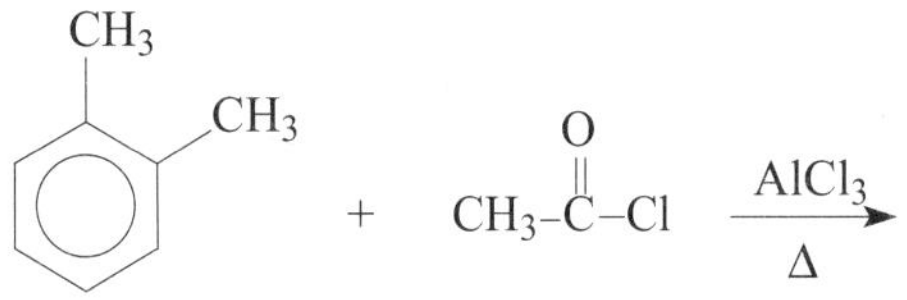 o-xylene + $CH_3-\overset{O}{\underset{\|}{C}}-Cl$ $\xrightarrow{AlCl_3, \Delta}$

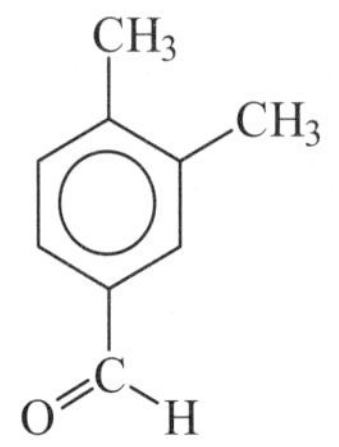

d) 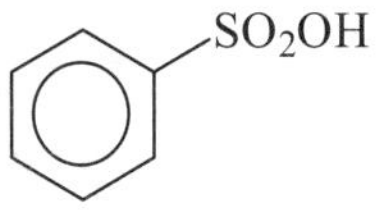 benzene + SO_3 $\xrightarrow{H_2SO_4}$

a) nitrobenzene (NO_2)

b) p-bromotoluene + o-bromotoluene

c) 3,4-dimethylbenzaldehyde

d) benzenesulfonic acid (SO_2OH)

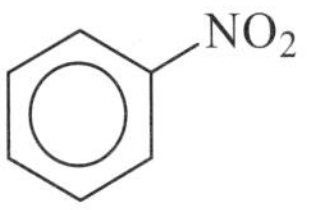

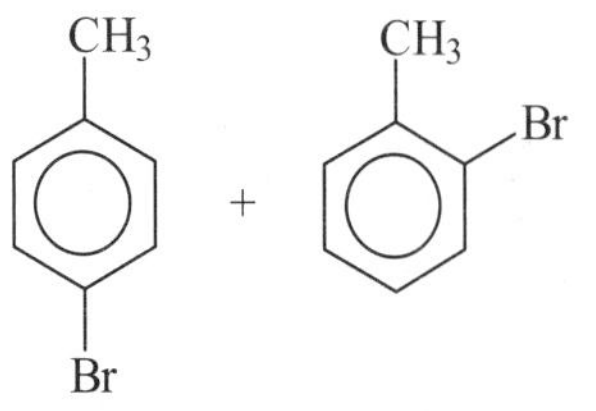

❏❏ **What kinds of conditions are usually required to hydrogenate Benzene?**

If a Nickel catalyst is used, then high temperatures (> 200 °C) and high pressures of Hydrogen (about 35 atm.) are required. Platinum catalysts, being more active, can effect the transformation at moderate pressures (2-3 atm) at room temperature.

❏❏ **In Electrophilic Aromatic Substitution, some substituents are said to be activating, and some are said to be deactivating. What does this mean?**

When there is an activating substituent on a phenyl ring, it causes the rate of that compound's substitution reactions to be faster than that of the analoguous reactions of Benzene. When there is a deactivating substituent on a phenyl ring, it causes the rate of that compound's substitution reactions to be slower than that of the analoguous reactions of Benzene.

❏❏ **What types of substituents are activating in Electrophilic Aromatic Substitution?**

Substituents that are electron donating by either induction or resonance stabilize the positive charge formed on the ring during substitution and therefore increase the rate.

❏❏ **What types of substituents are deactivating in Electrophilic Aromatic Substitution?**

Electron withdrawing substituents destabilize the positive charge formed on the ring and decrease the rate of substitution.

❏❏ **Activating substituents tend to direct ortho-para. Why is this?**

Because if a new substituent is added to either the ortho or the para position, one of the resonance structures will have the carbocation on the ring Carbon directly attached to the activating group. The activating (electron donating) group is best able to stabilize the carbocation (either by induction or by resonance) if it is in this position.

❏❏ **What's the easiest way to remember which groups direct ortho-para, and which direct meta?**

All Ortho-Para directors except for alkyl groups have a lone pair on the atom attached to the Benzene ring. No meta directing group has any lone pairs on the atom attached to the ring.

❏❏ **When Benzene is treated with Nitric Acid in Sulfuric Acid, the product is Nitrobenzene. What is the suspected electrophile in this reaction?**

The Nitronium Cation, NO_2^+.

❏❏ **How is a Nitronium Cation formed when Sulfuric Acid and Nitric Acid are mixed?**

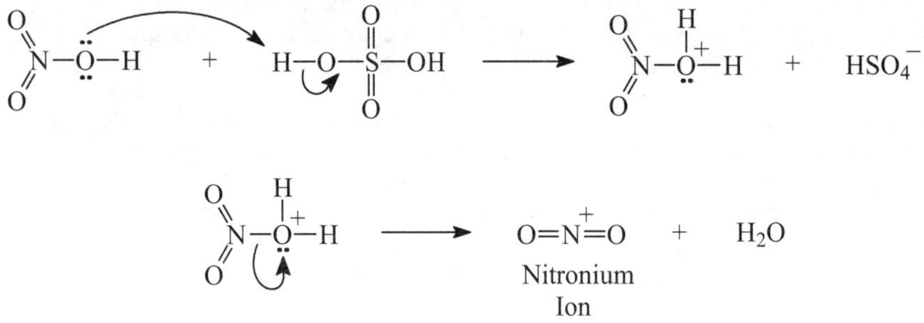

[Reaction scheme: HNO$_3$ + H$_2$SO$_4$ → protonated nitric acid + HSO$_4^-$]

[Second step: protonated nitric acid → O=N$^+$=O (Nitronium Ion) + H$_2$O]

❏❏ **Why do most deactivating substituents in Electrophilic Aromatic Substitution direct meta?**

Because substitution in the ortho a para positions place the carbocation directly adjacent to the electron withdrawing deactivating group. Substitution in the meta position does not.

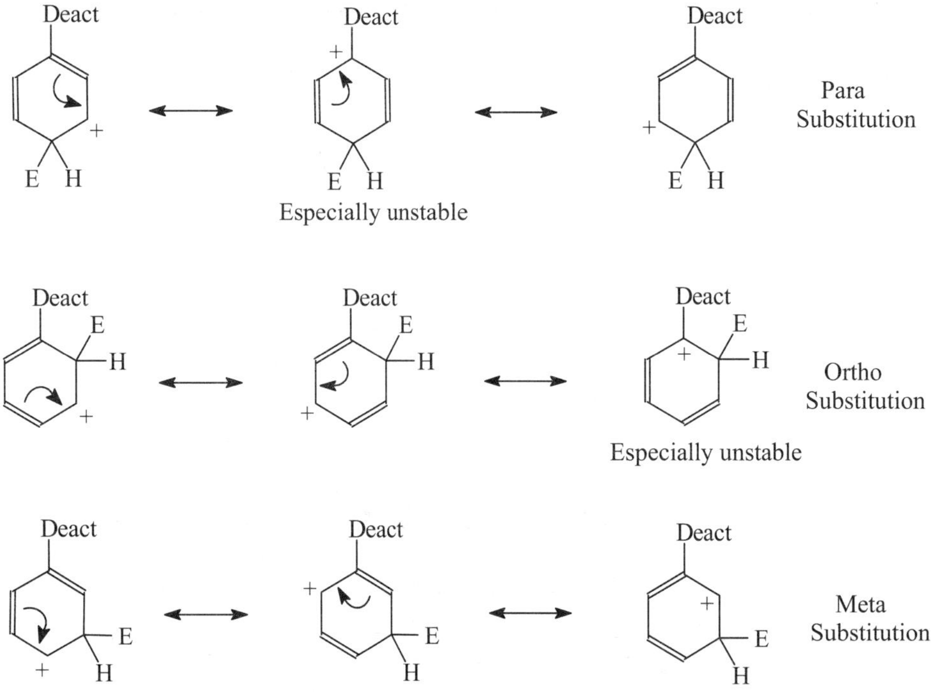

[Resonance structures for Para Substitution — middle structure labeled "Especially unstable"]

[Resonance structures for Ortho Substitution — right structure labeled "Especially unstable"]

[Resonance structures for Meta Substitution]

Carbocation never directly adjacent to the electron withdrawing deactivating group

❏❏ **When Alkenes react with electrophiles, the result is addition, but when Arenes react with electrophiles, the result is substitution. Why is this?**

Substitution preserves the aromaticity (energetically favorable), whereas addition would destroy it (energetically unfavorable.)

❏❏ **What is the Birch reduction?**

In the Birch reduction an Arene is treated with an alkali metal, usually Sodium, in liquid Ammonia in the presence of an Alcohol. The result is reduction of the Arene to a 1,4-Cyclohexadiene.

❏❏ **Though Halogen substituents are considered deactivating, they direct ortho-para. Why is this?**

Halogen atoms are deactivating due to their high electronegativities, but they also possess unshared pairs of electrons. These unshared pairs can help stabilize the carbocation if it is positioned on the Carbon directly adjacent to the Halogen. This can only occur in ortho or para substitution.

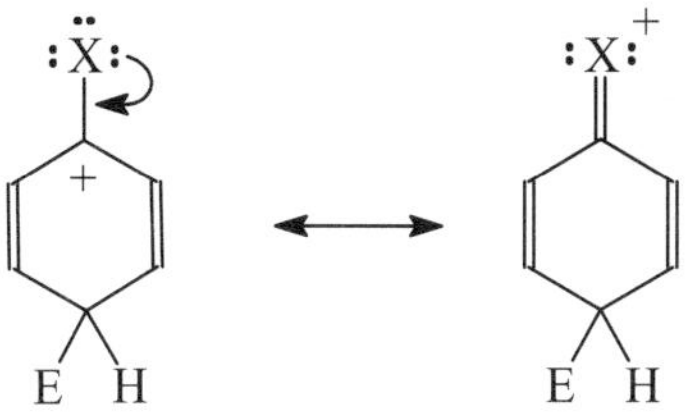

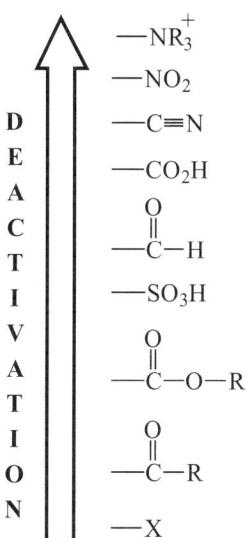

❏❏ **What are the common activating substituents in Electrophilic Aromatic Substitution? Arrange them in order of relative strength.**

ACTIVATION ↑

—NH$_2$ —NHR   —NR$_2$

—OH

—OR

—NH—C(=O)—R

—C$_6$H$_5$ (phenyl)

—R

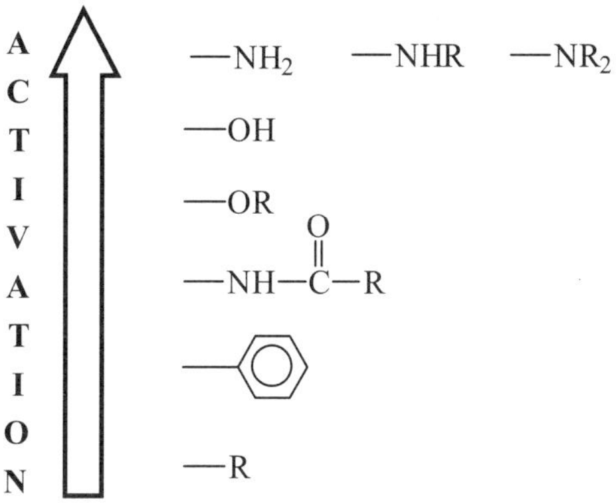

❏❏ **What are the common deactivating substituents in Electrophilic Aromatic Substitution? Arrange them in order of relative strength.**

DEACTIVATION ↑

—NR$_3^+$

—NO$_2$

—C≡N

—CO$_2$H

—C(=O)—H

—SO$_3$H

—C(=O)—O—R

—C(=O)—R

—X

❏❏ **Draw a reaction energy diagram for the addition of an electrophile to Benzene.**

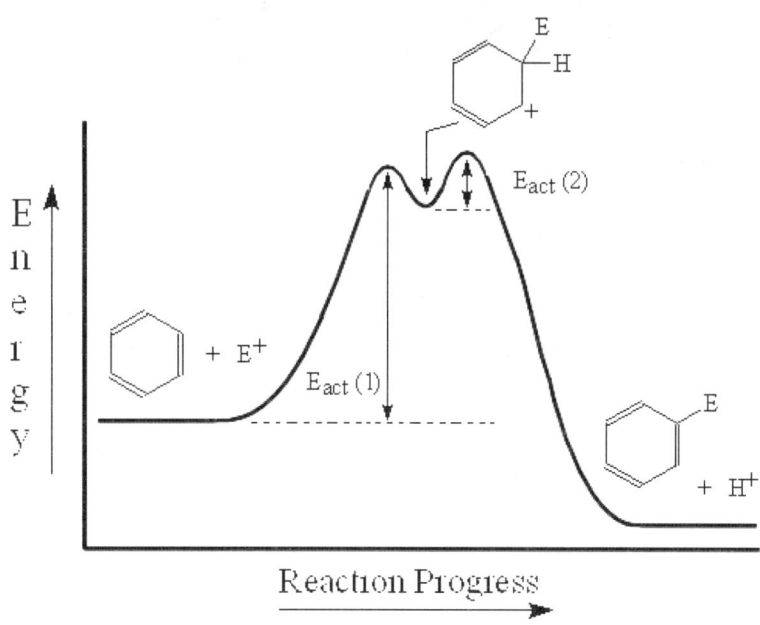

❏❏ **FeCl₃ is added to the reaction when carrying out the electophilic addition of Chlorine to Benzene. What is the role of the FeCl₃?**

The FeCl₃ forms a complex with the Chlorine to form an ion pair containing a highly electrophilic Chloronium ion. The Chloronium ion is the electrophile that attacks the Benzene ring to form the carbocation intermediate. In the final step of the reaction a proton is transferred to the other ion of the ion pair, $FeCl_4^-$, to give HCl and regenerate the FeCl₃ catalyst.

❏❏ **What is the catalyst most often used for a Friedel-Crafts reaction?**

Aluminum Trichloride.

❏❏ **What is the first step in the Friedel-Crafts alkylation of an Arene?**

Reaction of an Alkyl Halide with the Aluminum Trichloride to form an ion pair containing a carbocation electrophile.

❏❏ **What is a Friedel-Crafts acylation?**

It is an Electrophilic Aromatic Substitution reaction in which an Acid Chloride or Carboxylic Anhydride reacts with an aromatic compound in the presence of Aluminum Trichloride. The Acyl group becomes attached to the ring.

☐☐ **What are the drawbacks to Friedel-Crafts alkylation?**

Since the electrophile formed is a carbocation, elimination or rearrangement of the alkyl group can occur, or the electrophile can react with any stray nucleophiles present in the reaction mixture. In addition, the presence of any strongly deactivating groups on the phenyl ring will prevent reaction altogether. Also, it is sometimes difficult to limit the reaction to monoalkylation, since alkyl groups are activating substituents.

☐☐ **What advantage does a Friedel-Crafts acylation have over a Friedel-Crafts alkylation?**

In the alkylation reaction, the carbocation formed can possibly rearrange. Acyl cations, on the other hand, do not rearrange.

☐☐ **How could you best synthesize the following molecules from Benzene using a Friedel-Crafts acylation as the first step?**

a) 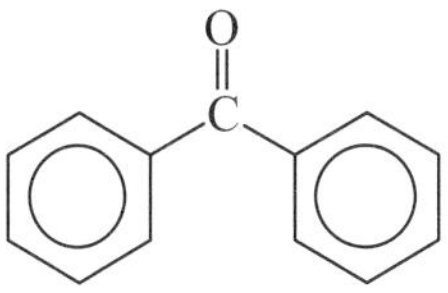 benzophenone   b) 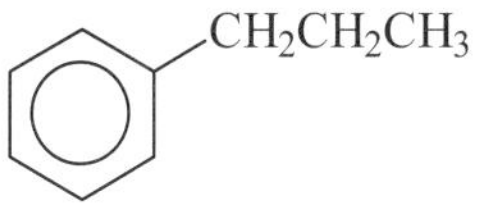 propylbenzene   c) 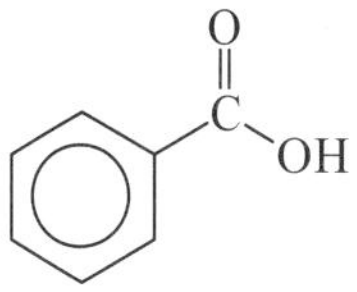 benzoic acid

a)  benzene + PhCOCl →(AlCl₃) benzophenone + HCl

b) 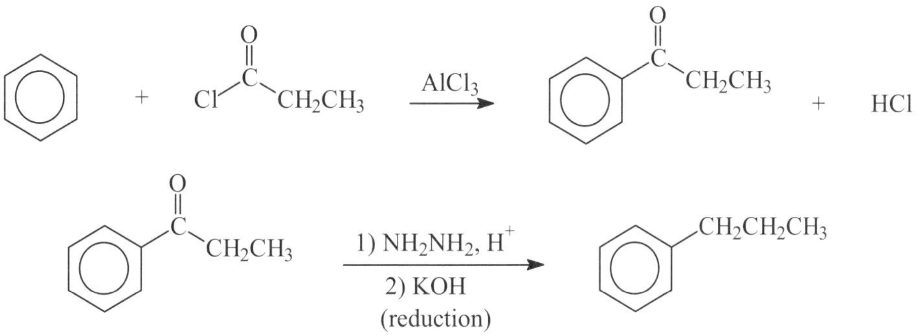 benzene + CH₃CH₂COCl →(AlCl₃) propiophenone + HCl

propiophenone →(1) NH₂NH₂, H⁺; 2) KOH (reduction)) propylbenzene

c) 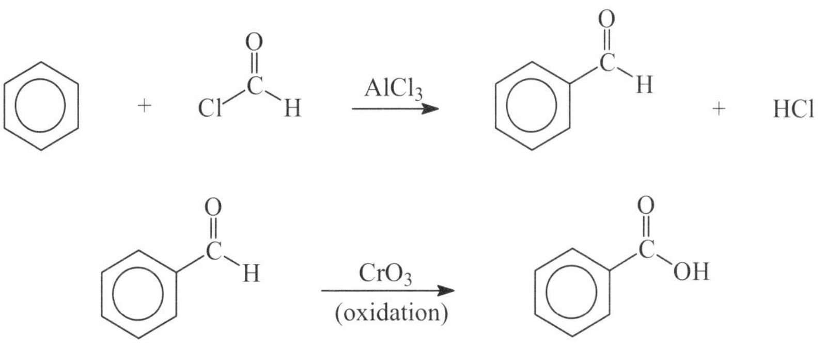 benzene + HCOCl →(AlCl₃) benzaldehyde + HCl

benzaldehyde →(CrO₃ (oxidation)) benzoic acid

☐☐ **Identify the following substituents as activating or deactivating in Electrophilic Aromatic Substitution, and indicate whether they direct meta or ortho-para.**

a) —CN   b) —NO$_2$ c) —OCH$_3$ d) —F   e) —NH—C(=O)—CH$_3$

(a) Deactivating, meta directing, (b) deactivating, meta directing, (c) activating, ortho-para directing, (d) deactivating, ortho-para directing, (e) activating, ortho-para directing

☐☐ **Hydroxyl groups and alkoxy groups are very electronegative, yet they are strong activators and direct ortho-para. Why is this?**

Because of the electronegativity of the Oxygen, hydroxyl and alkoxy substituents are electron withdrawing by the inductive effect, but this is overcome by a much greater electron donating resonance effect. The unshared pairs of electrons on the Oxygen can be donated to stabilize the carbocation formed on the phenyl ring if substitution occurs in the ortho or para positions.

Ortho Substitution   Para Substitution

Resonance stabilized carbocations

☐☐ **An Amide group is deactivating when attached to the Benzene ring through the Carbon, but activating when attached through the Nitrogen. How can this be explained?**

When the Amide group is attached to the ring through the carbonyl Carbon, the partial positive charge on the Carbon acts as a deactivating influence. When the Nitrogen of the Amide is attached to the ring, however, the unshared pair of electrons on the Nitrogen activates the ring through resonance effects.

Activating   Deactivating

☐☐ **Reaction of Pyridine with SO$_3$ in H$_2$SO$_4$ is much slower than the analogous reaction of Benzene. Why is this?**

Because in H$_2$SO$_4$, the Nitrogen of Pyridine is protonated. The positive charge on the ring deactivates it to substitution.

☐☐ **What effect do multiple substituents have on the site of further substitution on an aromatic ring?**

The directing effects of multiple substituents can either reinforce one another, or conflict. If they reinforce one another at any position, then that is where substitution will occur. If they conflict, then it is the more activating substituent that controls the site of further substitution.

☐☐ **Pyridine and Pyrrole are both weak bases, but Pyrrole is much weaker. Why is this?**

The lone pair on the Nitrogen of Pyrrole is part of the six electrons that make Pyrrole aromatic. In order for Pyrrole to act as a base, it must destroy the aromaticity of the ring by donating these electrons. This is not an energetically favorable thing to do. In Pyridine, on the other hand, the lone pair on the Nitrogen is completely independent from the aromatic system. Therefore Pyridine is free to donate the lone pair on the Nitrogen while remaining aromatic.

❏❏ **Where would nitration occur on the following molecules if they were treated with HNO₃/H₂SO₄?**

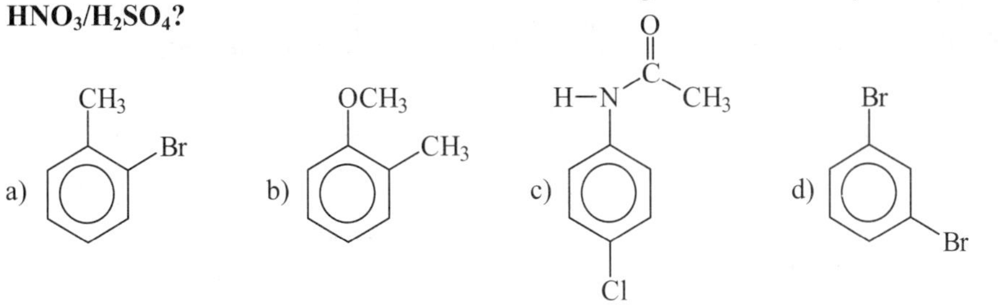

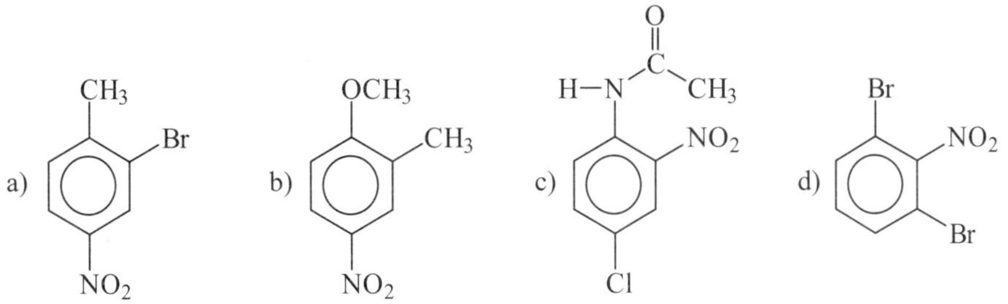

a) 2-bromo-1-methylbenzene b) 2-methoxy-1-methylbenzene (OCH₃ top, CH₃ adjacent) c) N-(4-chlorophenyl)acetamide d) 1,3-dibromobenzene

a) nitration para to CH₃ (NO₂ at position 4, opposite CH₃)
b) nitration para to OCH₃ (NO₂ at position 4)
c) nitration ortho to NHC(O)CH₃ (NO₂ adjacent to N, with Cl para)
d) nitration between the two Br groups (NO₂ at position 2)

❏❏ **[10] Annulene contains a total of 10 electrons in the π system (4n + 2, n = 2) and should be aromatic. The compound has, however, has been found to be non-aromatic. Why is this?**

If [10] Annulene were aromatic, it would have to be planar. Such a planar structure would force the two Hydrogens that face into the ring to lie almost on top of one another. The steric repulsion between these two Hydrogens forces [10] Annulene to be nonplanar, and therefore non-aromatic.

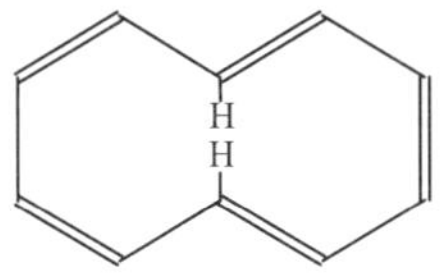

❏❏ **What is Kolbe carboxylation?**

It is a the reaction of a Phenoxide ion with Carbon Dioxide to give a Carboxylic Acid salt.

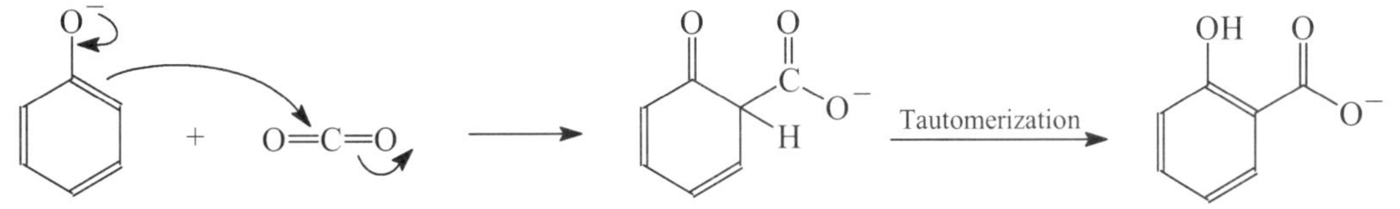

Phenoxide + $O=C=O$ → cyclohexadienone intermediate → (Tautomerization) → salicylate (2-hydroxybenzoate) anion

❏❏ **Benzene itself is very resistant to oxidation, but compounds such as Phenol and Catechol may be oxidized with Potassium Dichromate in Sulfuric Acid. What are the products of such oxidations?**

Quinones. Phenol produces *p*-Quinone and Catechol give *o*-Quinone

p-Quinone 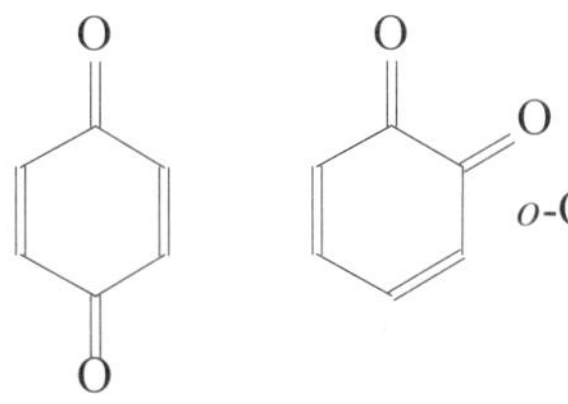   *o*-Quinone

❐❐ **What is the most interesting chemical property of Quinones?**

They are very easily reduced to Benzenediols. This reduction plays a part in a number of biological processes.

❐❐ **The benzylic position is particularly susceptible to reaction in aromatic compounds. Why is this?**

Cations and radicals produced at the Benzylic position is resonance stabilized and therefore easier to produce. Formation of such intermediates is the first step in many reactions.

❐❐ **Toluene, Ethylbenzene and Isopropylbenzene are all oxidized to Benzoic Acid on treatment with Chromic Acid.** *Tert*-**Butylbenzene, however, is unreactive. Why is this?**

Apparently a Benzylic Hydrogen is necessary to oxidize an alkyl substituent. *Tert*-Butylbenzene has no Benzylic Hydrogens.

❐❐ **What is An Aryl Diazonium salts and how is it produced?**

An Aryl Diazonium salt is a compound in which a diazonium group is attached to an aromatic ring. Aryl Diazonium salts are made by treating a primary Aryl Amine (such as Aniline) with Sodium Nitrite and acid at 0 - 5 °C.

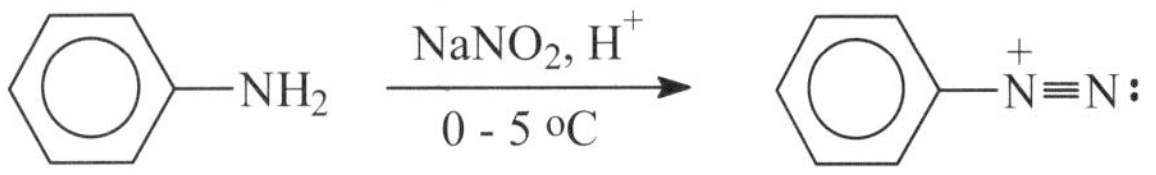

❐❐ **What is the product formed when Sodium Nitrite is acidified?**

Actually, a number of molecules that act as nitrosating agents are formed, but organic chemists tend to think in terms of only one of them, nitrous acid HONO. All of these species, however, act as a source of nitrosyl cation.

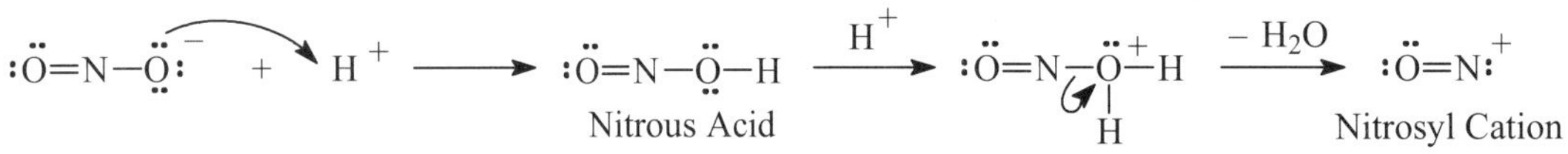

❐❐ **What is the mechanism for the formation of an Aryl Diazonium salt? Use Aniline as the starting material for your example.**

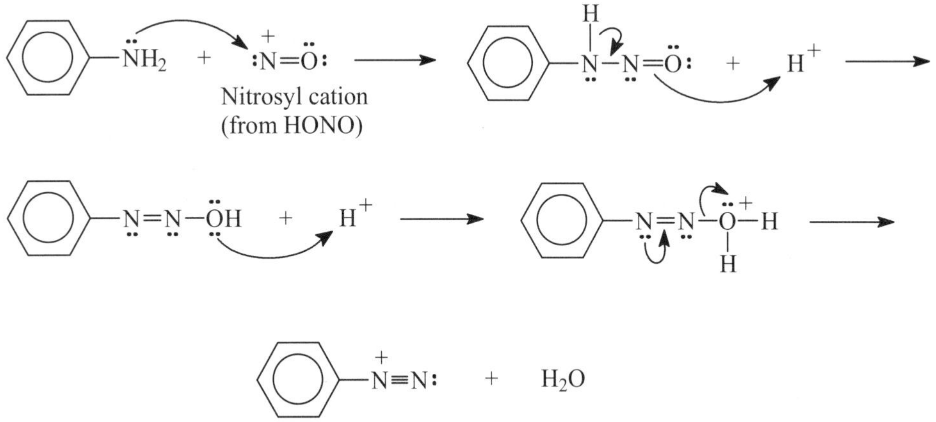

❏❏ **What are Aryl Diazonium salts used for?**

Aryl Diazonium salts are used to prepare a wide variety of rings substituted aromatic compounds by substitution of the Diazonium group. The Diazonium group is a good leaving group because it is expelled as a Nitrogen molecule.

❏❏ **What reagents would be needed to convert Benzenediazonium Chloride to each of the following molecules?**

a) Ph—OH   b) Ph—CN   c) Ph—Br   d) Ph

(a) H_2O, (b) CuCN, (c) CuBr, (d) H_3PO_2 or Ethanol

❏❏ **How are aryl fluorides prepared by the Schiemann reaction?**

An Aryl Diazonium salt is prepared by treating a primary Aryl Amine with Sodium Nitrite and acid to give the Diazonium salt. Tetrafluoroboric acid or Sodium Tetrafluoroborate is added to the mixture and the Diazonium Tetrafluoroborate salt precipitates out. The salt is collected and dried, and upon heating decomposes to give an Aryl Fluoride, Nitrogen gas, and Boron Trifluoride.

❏❏ **How could you prepare Benzoic Acid from Aniline using a Diazonium Salt as an intermediate?**

Convert the Aniline to a Diazonium Salt using Sodium Nitrite/acid at 0 - 5 °C. Then treat the Diazonium salt with CuCN to give Benzonitrile (Cyanobenzene). Hydrolysis of the Cyano group in aqueous acid will give Benzoic Acid.

❏❏ **What are the two major strengths of Diazonium Salt chemistry?**

(1) It is easy to introduce substituents to an aromatic ring, which are difficult to obtain by any other method.
(2) Compounds that have substituent patterns not directly available by Electrophilic Aromatic Substitution may be obtained.

❏❏ **What are Azo compounds?**

Azo compounds are compounds that contain an Azo (-N=N-) functionality. Often the azo group joins two aryl groups.

Ar—N=N—Ar

❏❏ **What is an Azo coupling reaction?**

In an azo coupling reaction a Diazonium salt is reacted with an aryl group possessing a strong activating group (such as OH or NR_2.) The product is an azo compound in which the two aryl groups are joined by an azo functionality.

❏❏ **In what capacity does the Diazonium Ion act in an Azo coupling reaction?**

As an electrophile.

❏❏ **What is the mechanism of an Azo coupling reaction?**

Act is a strongly activating group

❏❏ **What components were used to prepare the following azo compounds from a Diazonium salt?**

a) 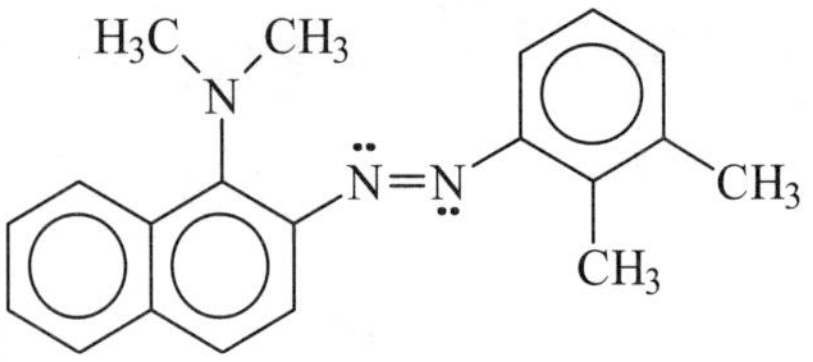 [N,N-dimethylamino naphthalene coupled to 2,3-dimethylphenyl via N=N]

b) 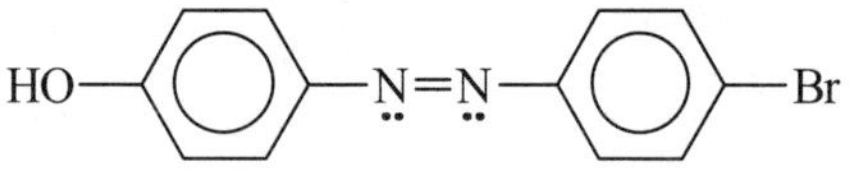 HO—C₆H₄—N=N—C₆H₄—Br

c) 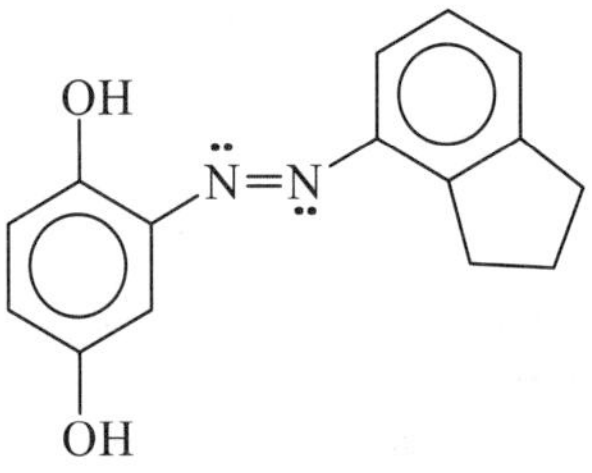 [2,5-dihydroxyphenyl coupled via N=N to indanyl]

d) 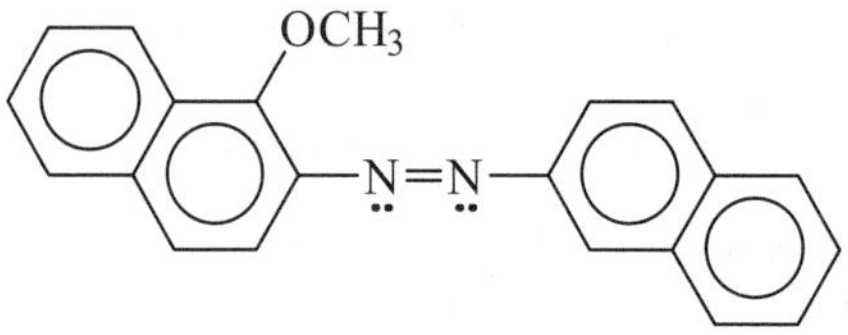 [2-methoxynaphthyl coupled via N=N to naphthyl]

a) 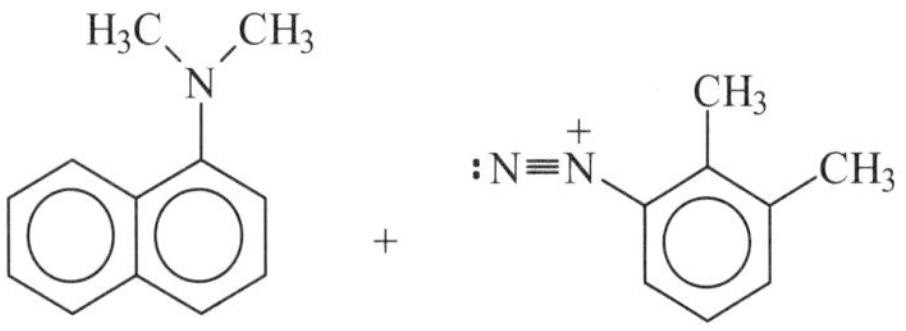 [1-(N,N-dimethylamino)naphthalene] + :N≡N⁺—[2,3-dimethylphenyl]

b) 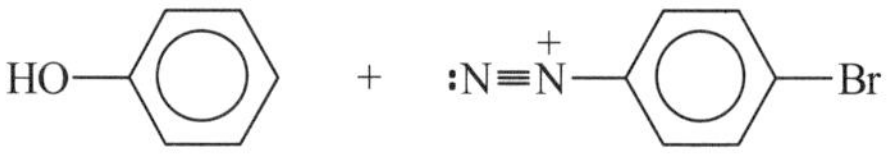 HO—C₆H₄—H + :N≡N⁺—C₆H₄—Br

a) 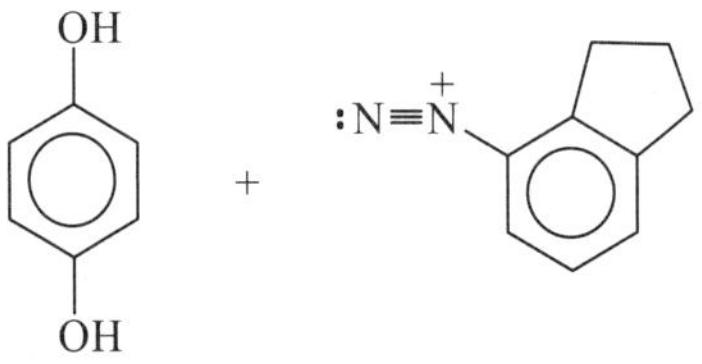 [1,4-dihydroxybenzene] + :N≡N⁺—[indanyl]

b) 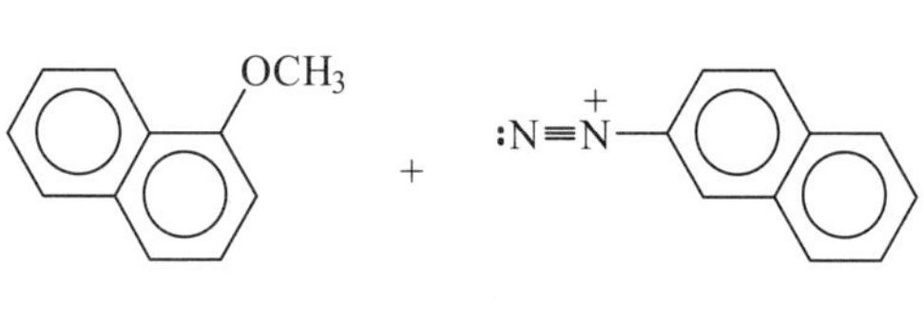 [2-methoxynaphthalene] + :N≡N⁺—[naphthyl]

❏❏ **What are Diaryl Azo compounds most commonly used for?**

Diaryl Azo compounds are most commonly used as dyes. The compounds are often highly colored and their color varies with the nature of the aryl groups.

❏❏ **What is Nucleophilic Aromatic Substitution?**

Nucleophilic aromatic substitution is a reaction in which a leaving group on an aromatic ring (typically a Halogen) is replaced as a substituent by a nucleophile.

❏❏ **What kind of kinetics are observed when a nucleophile such as Sodium Methoxide is reacted with a Nitro substituted aryl halide?**

The observed kinetics are second order, with the rate law being:

Rate = k [aryl halide] [nucleophile].

❏❏ **What structural features of the Aryl Halide can be changed to increase the rate of nucleophilic aromatic substitution?**

Increased nitro substitution in the positions ortho and para to the Halogen will increase the rate dramatically. The rate of substitution also changes with the Halogen being replaced, with the rate being F > Cl > Br > I.

❏❏ **What is the mechanism of Nucleophilic Substitution on a Nitrated Aryl Halide?**

Nucleophilic Aromatic Substitution on a Nitro substituted Aryl Halide follows a two step addition-elimination mechanism.

❏❏ **What is the rate determining step of this mechanism?**

The addition step. This is not surprising, since the aromaticity of the ring is temporarily destroyed in this step. The elimination step is quite rapid because the aromaticity is restored.

❏❏ **Why does increasing the number of Nitro substituents ortho and para to the Halogen increase the rate of this reaction?**

Nitro groups are strongly electron withdrawing. When they are ortho or para to the Halogen position they are capable of stabilizing the negative charge formed in the rate-determining step. Stabilizing the charge results in a more stable anionic intermediate and therefore a lower activation energy.

❏❏ **Fluoride is usually a lousy leaving group, yet Nitro substituted Aryl Fluorides react more rapidly than the other Aryl Halides in Electrophilic Aromatic Substitution reactions. How is this possible?**

The step in which the Halogen is expelled in the Electrophilic Aromatic Substitution of Nitro Aryl Halides is not the rate determining step. Therefore the leaving group ability of the Halogen is irrelevant. The rate determining step is addition of the nucleophile to give an anionic intermediate. Anything that helps to stabilize that anion will increase the rate of the reaction. Fluoride is the most electronegative of the Halogens, and its electron withdrawing ability helps to stabilize the anion formed. Since it can do this better than any of the other Halogens, Nitro Aryl Fluorides react more rapidly than the other Nitro Aryl Halides.

❏❏ **Very strong bases such as Sodium Amide readily react with Aryl Halides even when they are not activated by Nitro groups. What kind of intermediate has been proposed in these reactions?**

A Benzyne intermediate.

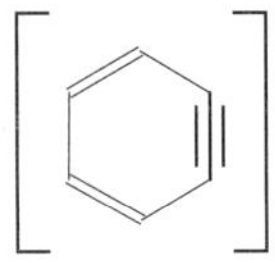

❏❏ **How are these reactions usually carried out?**

An Aryl Halide is reacted with Sodium or Potassium Amide in liquid Ammonia at low temperatures (about -33 °C).

❏❏ **What is the mechanism of these Nucleophilic Substitution Reactions?**

In this case the proposed mechanism is a two step elimination-addition mechanism:

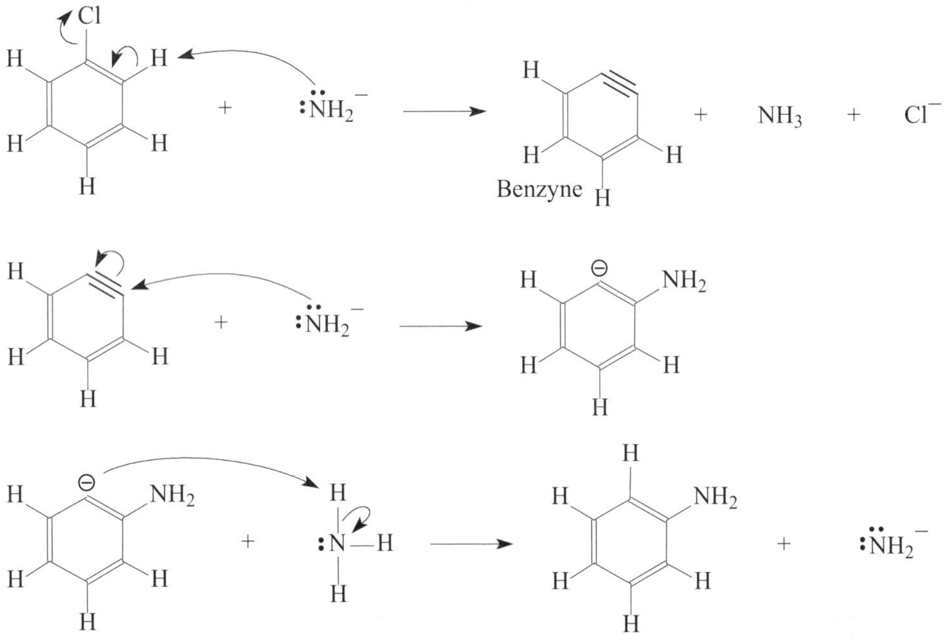

❏❏ **Benzyne is a very highly reactive intermediate. Why is this?**

The Carbons involved in the triple bond of Benzyne are sp hybridized and should have bond angles of 180 degrees. The ring structure of Benzyne, however, constricts them to 60 degrees bond angles. This creates a great amount of strain in the molecule, making it highly reactive.

❏❏ **What is the Regiochemistry of substitution in these reactions?**

Substitution may occur at either the position originally occupied by the Halogen, or at a position ortho to it. This is because the Benzyne intermediate renders these two positions chemically equivalent.

❏ ❏ **Though they have no nitro substituents, both Hexafluorobenzene and 2-Chloropyridine undergo rapid reaction with nucleophiles. Why is this?**

In the case of Hexafluorobenzene the combined electron withdrawing ability of the six Fluorines compensates for the lack of a nitro substituent. In 2-Chloropyridine, the anionic intermediate is stabilized by a resonance form in which the negative charge resides on the (electronegative) Nitrogen of the Pyridine ring.

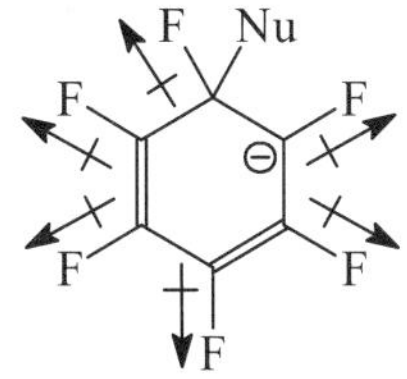

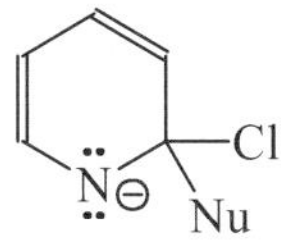

ALDEHYDE AND KETONE PEARLS

Organic chemists, bunglers, "students of gunk"...
Primo Levi, in *The Periodic Table,* translated by Raymond Rosenthal

❏❏ **What is the characteristic feature of a Ketone?**

A carbonyl group (C=O) with two alkyl groups attached to the carbonyl Carbon.

❏❏ **What is the characteristic feature of an Aldehyde?**

A carbonyl group (C=O) with at least one Hydrogen attached to the carbonyl Carbon. The other group may be alkyl, aryl, or a second Hydrogen.

❏❏ **What is the most typical reaction of Aldehydes and Ketones?**

Addition of a nucleophile to the carbonyl to give a tetrahedral Carbon intermediate, which may then react in a variety of ways.

❏❏ **How are Aldehydes named in IUPAC nomenclature?**

The longest continuous chain containing the Aldehyde group provides the parent name. The -e ending of the Alkane is replaced with -al. Substituents are indicated in the usual way, with their positions indicated by the number of the parent chain Carbon to which they are attached. Numbering of the chain begins at the end bearing the Aldehyde group. When two Aldehyde groups are present at either end of the chain, then the suffix -dial is used. When a formyl (CHO) group is attached to a ring, the ring is named, followed by the suffix -carbaldehyde.

❏❏ **How are Ketones named in IUPAC nomenclature?**

The longest continuous chain containing the Ketone group provides the parent name. The -e ending of the Alkane is replaced with -one. Substituents are indicated in the usual way, with their positions indicated by the number of the parent chain Carbon to which they are attached. Numbering of the chain begins at the end closest to the Ketone carbonyl, and the position of the carbonyl must be indicated with a number preceding the name of the parent chain. When two or more Ketone groups are present in the molecule, then the suffixes -dione, trione, etc. are used. IUPAC nomenclature also allows for the use of radicofunctional names, in which the groups attached to the carbonyl are each named as separate words, followed by the word ketone. The attached groups should be listed in alphabetical order.

❏❏ **What would be the IUPAC name for each of the following molecules?**

a) CH₃CH₂CH₂CH₂—C(=O)—H   b) [cyclohexyl]—CH₂—C(=O)—CH₃

c) [cyclopentanone structure]   d) [cyclopropyl]—C(=O)—H

(a) Pentanal, (b) 1-Cyclohexyl-2-propanone, (c) Cyclopentanone, (d) Cyclopropane carbaldehyde

❏❏ **What are the Radicofunctional names of the following compounds?**

a), 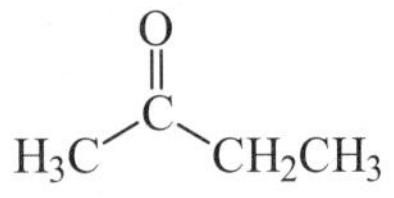 b), 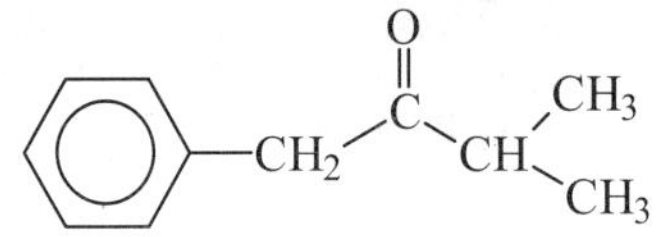 c), 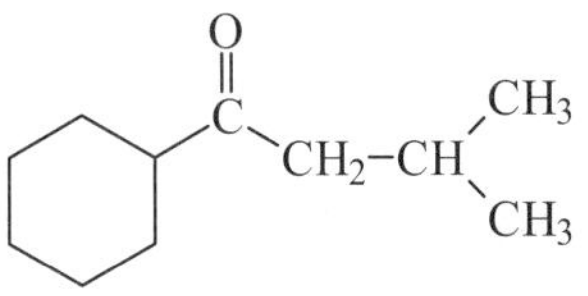 d) 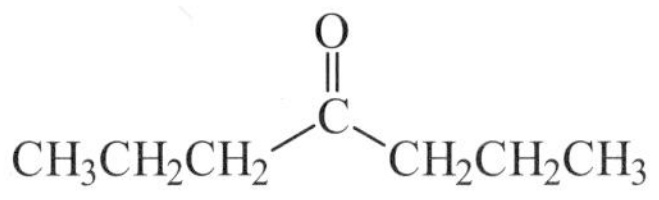 (structures shown)

(a) Methyl Ethyl Ketone, (b) Benzyl Isopropyl Ketone, (c) Cyclohexyl Isobutyl Ketone, (d) Dipropyl Ketone

❏❏ **What are the common names of the following compounds?**

a), 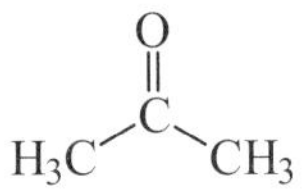 b), 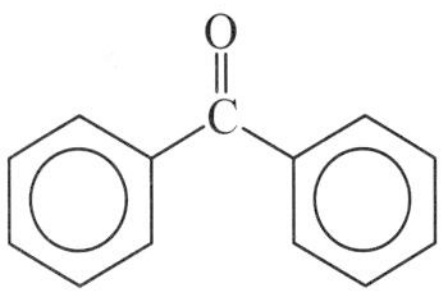 c), 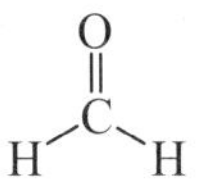 d), 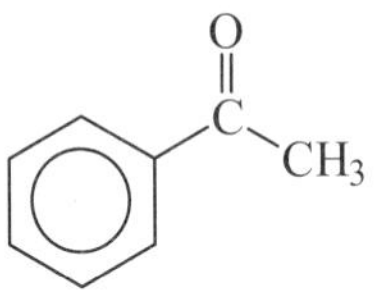 e), 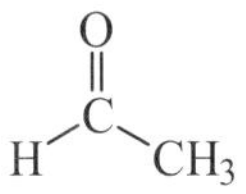 f) 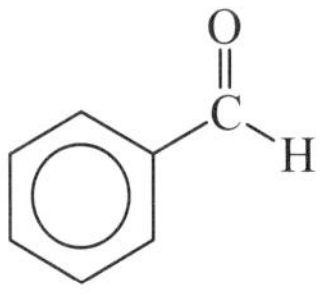 (structures shown)

(a) Acetone, (b) Benzophenone, (c) Formaldehyde, (d) Acetophenone, (e) Acetaldehyde, (f) Benzaldehyde

❏❏ **What prefix is used when the Oxygen of a Carbonyl must be named as a substituent?**

Oxo.

❏❏ **What is the geometry of a Carbonyl compound?**

The bond angles around the Carbonyl Carbon are close to 120 degrees, the ideal for an sp^2 hybridized Carbon. The Carbonyl Oxygen and Carbon, and the two atoms bonded to the Carbonyl all lie in a plane.

❏❏ **What is Hydroformylation?**

Hydroformylation is a reaction used (usually on an industrial scale) for the preparation of Aldehydes. It involves the reaction of Alkenes with Carbon Monoxide and Hydrogen gas in the presence of a Cobalt or Rhodium based catalyst. The reaction is also known as the "oxo process."

$$R-CH=CH_2 + CO + H_2 \xrightarrow{[Co(CO)_4]_2} RCH_2CH_2-\overset{\overset{O}{\|}}{C}-H$$

❏❏ **What is the direction of the polarity of a Carbonyl group?**

The Carbon end is slightly positive and the Oxygen end is slightly negative.

$$\overset{\longrightarrow}{\delta^+ \; C=O \; \delta^-}$$

❏❏ **Smaller Aldehydes and Ketones are quite soluble in water. Why is this?**

The polarity of the Carbonyl group is strong enough to make smaller Aldehydes and Ketones soluble in the highly polar aqueous environment.

❏❏ **The reaction of Grignard reagents with carbonyl compounds is a very good way to prepare a wide variety of Alcohols. What type of carbonyl compound would be needed to prepare each of the following types of Alcohols?**

a) Primary   b) Secondary   c) Tertiary

(a) Reacting a Grignard reagent with Formaldehyde will give a primary Alcohol. For example, reaction of Methyl Magnesium Bromide with Formaldehyde gives Ethanol.

$$\underset{H}{\overset{O}{\underset{\|}{H-C-H}}} \xrightarrow[\text{2) } H_3O^+]{\text{1) } CH_3^- \; MgBr^+} \underset{\text{a primary alcohol}}{CH_3CH_2OH}$$

(b) Reaction of Grignards with other Aldehydes gives secondary Alcohols, as in the reaction between Methyl grignard and Acetaldehyde to give 2-Propanol.

$$\underset{H_3C}{\overset{O}{\underset{\|}{C}}}\underset{H}{} \xrightarrow[\text{2) } H_3O^+]{\text{1) } CH_3^- \; MgBr^+} \underset{\text{a secondary alcohol}}{H_3C-\underset{H}{\overset{OH}{\underset{|}{C}}}-CH_3}$$

(c) Reaction between Grignard reagents and Ketones yields tertiary Alcohols, as when Methyl grignard reacts with Acetone to give *t*-Butanol.

$$\underset{H_3C}{\overset{O}{\underset{\|}{C}}}_{CH_3} \xrightarrow[\text{2) } H_3O^+]{\text{1) } CH_3^- \; MgBr^+} \underset{\text{a tertiary alcohol}}{H_3C-\underset{CH_3}{\overset{OH}{\underset{|}{C}}}-CH_3}$$

❏❏ **What is the mechanism of the reaction between a Grignard reagent and a carbonyl compound? Use Acetone and Methyl Magnesium Bromide for your example.**

$$\underset{H_3C}{\overset{O}{\underset{\|}{C}}}_{CH_3} + CH_3-MgBr \longrightarrow H_3C-\underset{CH_3}{\overset{O^-}{\underset{|}{C}}}-CH_3 \xrightarrow{H_3O^+} H_3C-\underset{CH_3}{\overset{OH}{\underset{|}{C}}}-CH_3$$

◻◻ **How is it possible to prepare Carboxylic Acids using a Grignard reagent?**

React them with Carbon Dioxide. This is usually done by pouring a prepared Grignard reagent over dry ice. This gives the salt of the Carboxylic Acid, which can be converted to the free acid by treatment with HCl.

$$O=C=O \;+\; R^- MgBr^+ \longrightarrow R-\underset{O^-\; MgBr^+}{\overset{O}{\underset{\|}{C}}} \xrightarrow{H_3O^+} R-\underset{OH}{\overset{O}{\underset{\|}{C}}}$$

◻◻ **How could you prepare each of the following Alcohols using a Grignard?**

a) cyclopropyl–CH₂–OH

b) $CH_3CH_2-\underset{CH_2CH_3}{\overset{OH}{\underset{|}{C}}}-CH_2CH_3$

c) Ph–C(OH)(H)–CH₂CH₂CH₃

a) cyclopropyl–MgBr + H–CHO

b) CH₃CH₂–MgBr + CH₃CH₂–CO–CH₂CH₃

c) Ph–MgBr + H–CO–CH₂CH₂CH₃   or   Ph–CO–H + BrMg–CH₂CH₂CH₃

◻◻ **What are the advantages and disadvantages of using an Organolithium compound to prepare Alcohols from Aldehydes and Ketones?**

Organolithium compounds are more reactive than Grignards and usually give higher yields. However, they are so sensitive that they must be prepared and handled under an inert atmosphere such as Nitrogen or Argon.

◻◻ **The salts of terminal Alkynes can act as nucleophiles in reactions with Carbonyl compounds. What is the product is obtained in such a reaction?**

Treatment of an Aldehyde or Ketone with the salt of a terminal Alkyne (followed by acid work-up) gives a compound that contains both an Alcohol and an Alkyne functionality.

$$R-\overset{O}{\underset{\|}{C}}-R \;+\; R-C\equiv C^- \; Na^+ \longrightarrow R-C\equiv C-\underset{R}{\overset{R}{\underset{|}{\overset{|}{C}}}}-O^- \xrightarrow{H_3O^+} R-C\equiv C-\underset{R}{\overset{R}{\underset{|}{\overset{|}{C}}}}-OH$$

◻◻ **Why is this reaction synthetically very useful?**

The reaction creates a new C–C bond, which is always a useful synthetic process. In addition, depending on the structure of the Alkyne that was used, the triple bond present in the product can be hydrated to an Aldehyde or Ketone. This new carbonyl can then undergo further nucleophilic reactions to create even more complicated structures.

❐❐ **What is a Cyanohydrin and how is it formed?**

A Cyanohydrin is a molecule containing a Cyano group (–CN) and a Hydroxyl group (–OH) bonded to the same Carbon. They are formed by addition of HCN to the carbonyl of an Aldehyde or Ketone.

$$N{\equiv}C\underset{RR}{\overset{OH}{\diagdown C \diagup}} \quad \text{A Cyanohydrin}$$

❐❐ **What are the typical reaction conditions under which an Aldehyde or Ketone is converted to a Cyanohydrin?**

The reaction is usually carried out by treating the carbonyl compound with an aqueous solution of NaCN or KCN, which has been adjusted to a pH of 10. This gives a mixture of HCN and Cyanide ion (which is the nucleophile). HCN cannot be added directly to the carbonyl compound because Hydrogen Cyanide is such a weak acid that only trace amounts of CN⁻ exist in aqueous solution.

❐❐ **What is the mechanism for the formation of a Cyanohydrin?**

The reaction is formally an addition of HCN to the carbonyl, and is catalyzed by cyanide ion.

❐❐ **What are the limitations of the Cyanohydrin formation reaction?**

The reaction is an equilibrium. For Aldehydes and unhindered Aliphatic Ketones, the equilibrium favors the Cyanohydrin. But for most Aryl Ketones, and for sterically hindered Aliphatic Ketones, the equilibrium favors the starting materials.

❐❐ **What is a Phosphorus Ylide?**

An Ylide is a neutral molecule in which two oppositely charged atoms, each with complete octets, are bonded directly to one another. A Phosphorus Ylide has the structure:

$R_3P^{+}{-}\overset{..}{C}^{-}\diagdown$ R is often a phenyl group

❐❐ **How are Phosphorus Ylides formed?**

Usually by treatment of a primary or secondary Alkyl Halide (X = I, Cl, or Br) with a trivalent phosphorus compound (such as Triphenylphosphine), followed by reaction of the resulting salt with a strong base such as Butyl Lithium.

$$CH_3I \ + \ Ph_3P: \ \longrightarrow \ Ph_3\overset{\oplus}{P}{-}CH_3 \ \ I^{\ominus} \ \xrightarrow{\text{Butyl Lithium}} \ Ph_3\overset{\oplus}{P}{-}\overset{..}{\underset{\ominus}{C}}H_2$$

a Phosphorus Ylide

❏❏ **When Phosphorus Ylides are reacted with Ketones or Aldehydes, an interemediate, known as an Oxaphosphetane, can be isolated at low temperatures. What is the structure of this intermediate?**

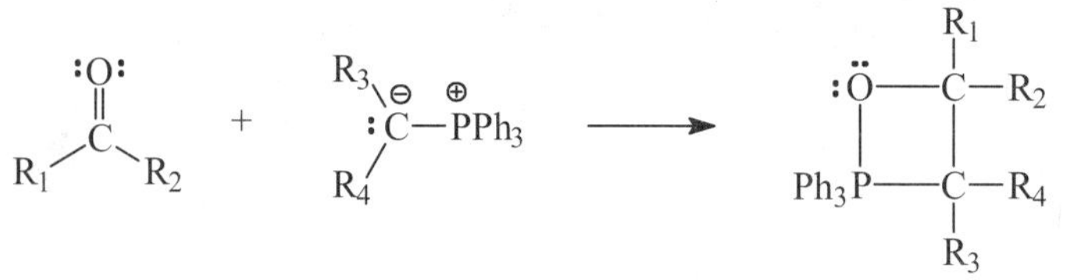

If the Ylide is made using Triphenylphosphine, the Oxaphosphetane will have the structure shown.

An Oxaphosphetane

❏❏ **What is the mechanism of the Wittig reaction?**

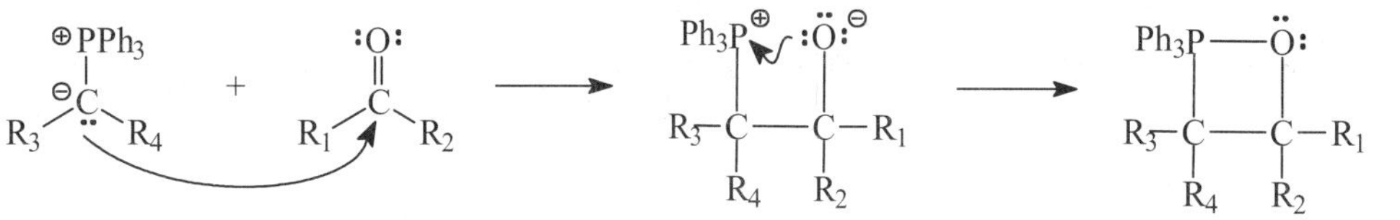

a Betaine

An Oxaphosphetane

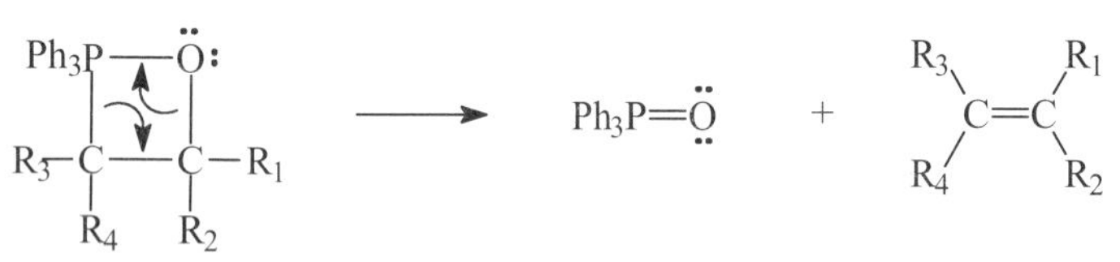

❏❏ **Why is the Wittig reaction such an attractive method for the preparation of Alkenes?**

There are three reasons: (1) the reaction forms a C–C bond, allowing elaborate Carbon skeletons to be constructed, (2) It is tolerant of many other functional groups, including double and triple bonds, Alcohols, Ethers, and Esters, and (3) It is very regioselective; the double bond forms between the carbonyl Carbon and the negatively charged Carbon of the Ylide.

❏❏ **Triphenylphosphine is the reagent of choice for the formation of the Wittig reagent. Why is this?**

Triphenylphosphine is a solid at room temperature (m. pt. = 81 °C), making it easy to handle and measure. This can be extremely important because Phosphines are generally toxic.

❏❏ **Match the following terms**

1) Cyanohydrin   a) Ketones

1) Ylide Precursor   b) Aldehydes

3) Oxidized by CrO3   c) $Ph_3P^+CH_3$ I^-

4) Oxidized by peroxy acids   d) Schiff base

5) Carbonyl + primary Amine   e) $R_2C(OH)CN$

(1) e, (2) c, (3) b, (4) a, (5) d.

❏❏ **How could each of the following molecules be prepared using a Wittig reaction?**

a) $CH_3CH_2CH=C(CH_3)-CH_3$ (with CH₃ branch)

b) cyclohexane with =CHCH₃ substituent

c) $(H_3C)_2C=C(CH_3)_2$

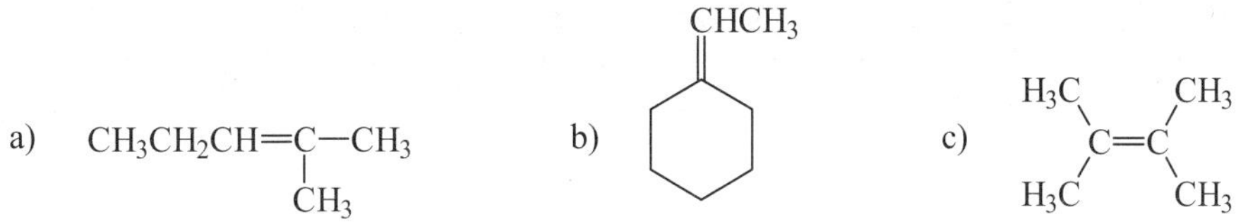

Since every double bond has two ends, there are typically two possible ways to make any Alkene using a Wittig reaction. The only real exception is in the case of highly symmetrical molecule.

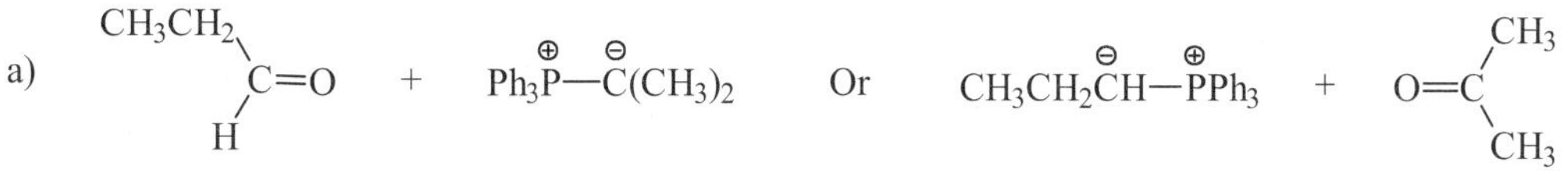

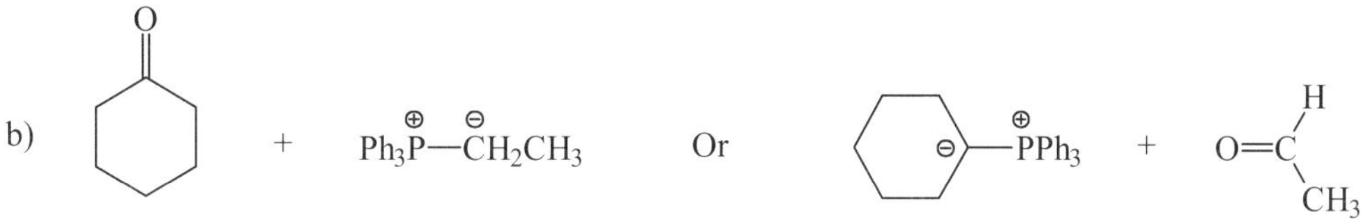

a) $CH_3CH_2\text{-}CH\text{=}O$ (propanal) + $Ph_3\overset{\oplus}{P}-\overset{\ominus}{C}(CH_3)_2$ Or $CH_3CH_2\overset{\ominus}{C}H-\overset{\oplus}{P}Ph_3$ + $O=C(CH_3)_2$

b) cyclohexanone + $Ph_3\overset{\oplus}{P}-\overset{\ominus}{C}H_2CH_3$ Or   cyclohexyl-$\overset{\ominus}{}$-$\overset{\oplus}{P}Ph_3$ + $O=CH(CH_3)$

c) $(H_3C)_2C=O$ + $Ph_3\overset{\oplus}{P}-\overset{\ominus}{C}(CH_3)_2$

❏❏ **What is keto-enol tautomerism?**

Keto-enol tautomerism is an equilibrium between a Ketone or Aldehyde and it's isomeric enol form. The transformation is accomplished by a formal movement of a proton.

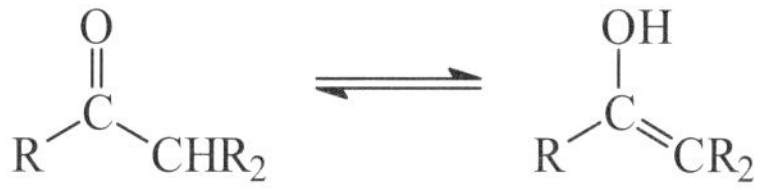

$$\underset{\text{Keto form}}{R-\overset{O}{\underset{\|}{C}}-CHR_2} \rightleftharpoons \underset{\text{Enol form}}{R-\overset{OH}{\underset{|}{C}}=CR_2}$$

❏❏ **Which is the favored form in the keto-enol equilibrium?**

The keto form is usually the favored isomer at equilibrium.

❏❏ **In what type of molecule might the enol form be more stable?**

The enol form is the predominant form in β-diketones. This is because of the increased stability of the enol form resulting from the conjugation of the double bond with a carbonyl. In addition, the enol form is set up well for hydrogen bonding between the enol Hydrogen and the Oxygen of the other carbonyl. This contributes additional stability.

(diagram: acetylacetone keto form ⇌ enol form with intramolecular H-bond)

☐☐ Why are the α Hydrogens of an Aldehyde or Ketone Acidic?

The anion resulting from the deprotonation of an Aldehyde or Ketone in the α position is resonance stabilized and therefore fairly easy to form. This translates into a greater ability to lose an α proton, hence higher acidity than the typical proton on an sp^3 hybridized Carbon.

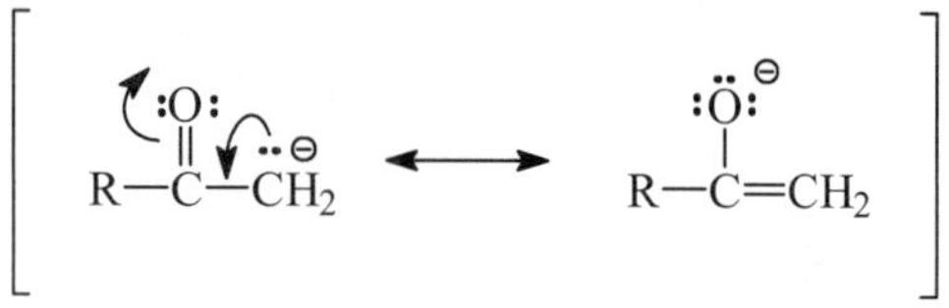

$$\left[\text{R}-\underset{\underset{\text{CH}_2}{\|}}{\overset{:\text{O}:}{\text{C}}}^{\ominus} \quad \longleftrightarrow \quad \text{R}-\underset{\text{CH}_2}{\overset{:\ddot{\text{O}}:^{\ominus}}{\text{C}}}= \right]$$

☐☐ What is an Enolate?

An enolate or enolate anion is the anion resulting from the loss of a Hydrogen from the alpha position of a carbonyl group. It is so named because it is the anion (deprotonated form) of an enol.

☐☐ What reagents could be used to perform each of the following transformations?

a)  2-butene → acetaldehyde

b) 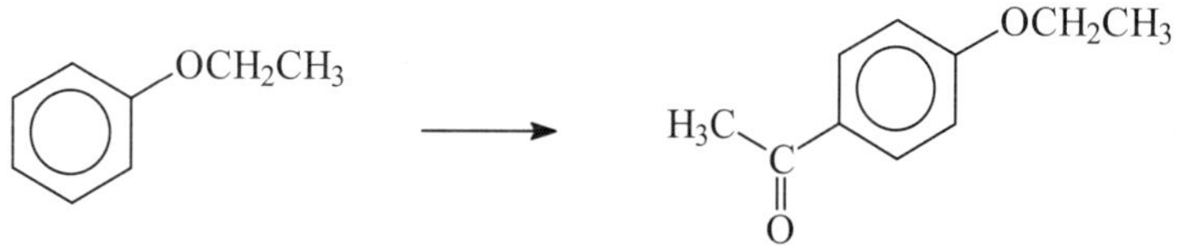 phenyl ethyl ether → 4'-ethoxyacetophenone

c) 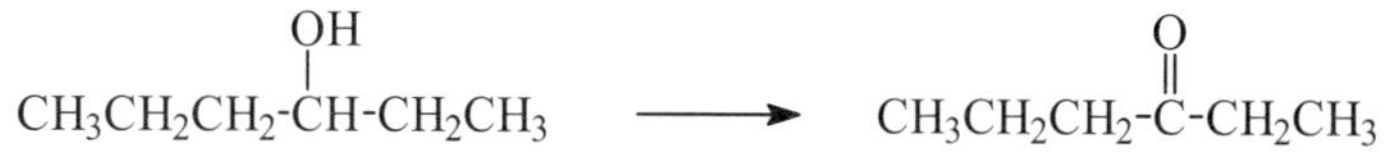 3-hexanol → 3-hexanone

d)  cyclohexylmethanol → cyclohexanecarbaldehyde

(a) O_3 followed by treatment with Zn / H_2O, (b) Acetyl Chloride (CH_3COCl) in the presence of $AlCl_3$, (c) Chromate or Dichromate (Cr VI), (d) Pyridinium Chlorochromate (PCC)

❑❑ **What would be the primary organic product(s) of the following reactions?**

a) 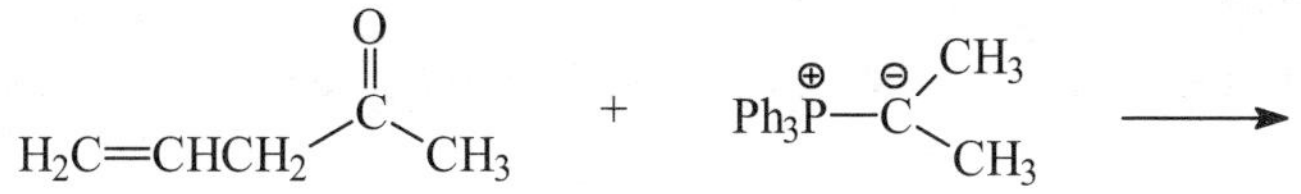 H₂C=CHCH₂C(=O)CH₃ + Ph₃P⁺—C⁻(CH₃)(CH₃) ⟶

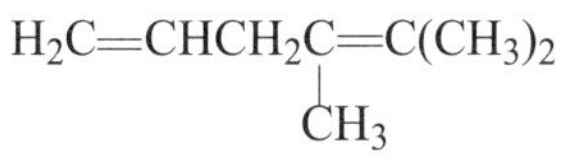

b) 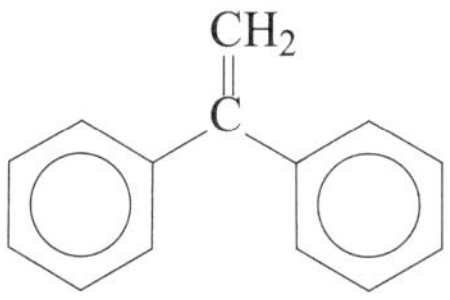 Ph₂C=O + Ph₃P⁺—C⁻H(H) ⟶

c) CH₃CH₂C(=O)CH₃ + Ph₃P⁺—C⁻(CH₃)(CH₂CH₃) ⟶

a) H₂C=CHCH₂C(CH₃)=C(CH₃)₂   b) Ph₂C=CH₂

c)
CH₃CH₂\\C=C//CH₃\\CH₂CH₃ /H₃C   and / or   CH₃CH₂\\C=C//CH₂CH₃\\CH₃ /H₃C

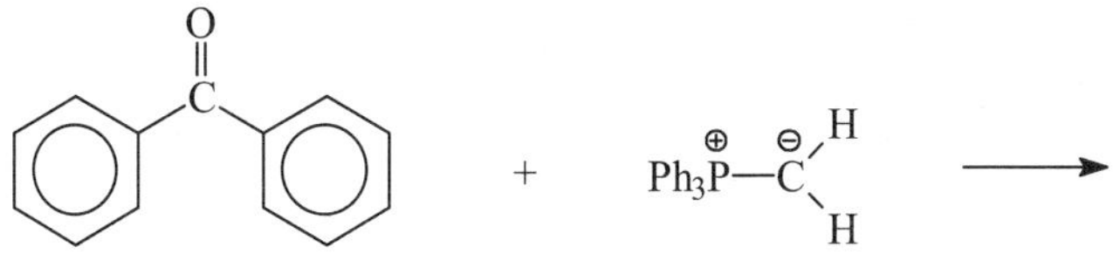

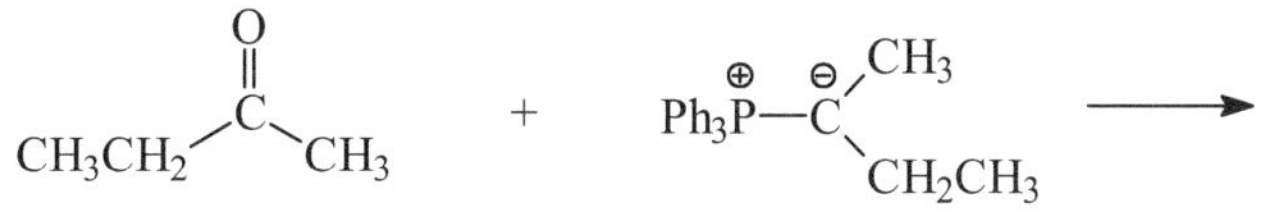

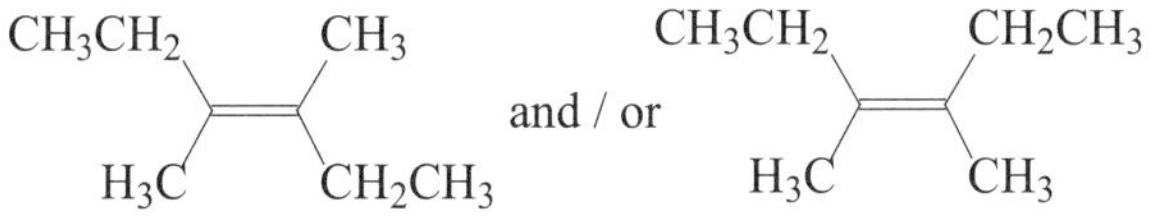

Ratio of isomers would depend on reaction conditions

❑❑ **What is the mechanism for the formation of an imine?**

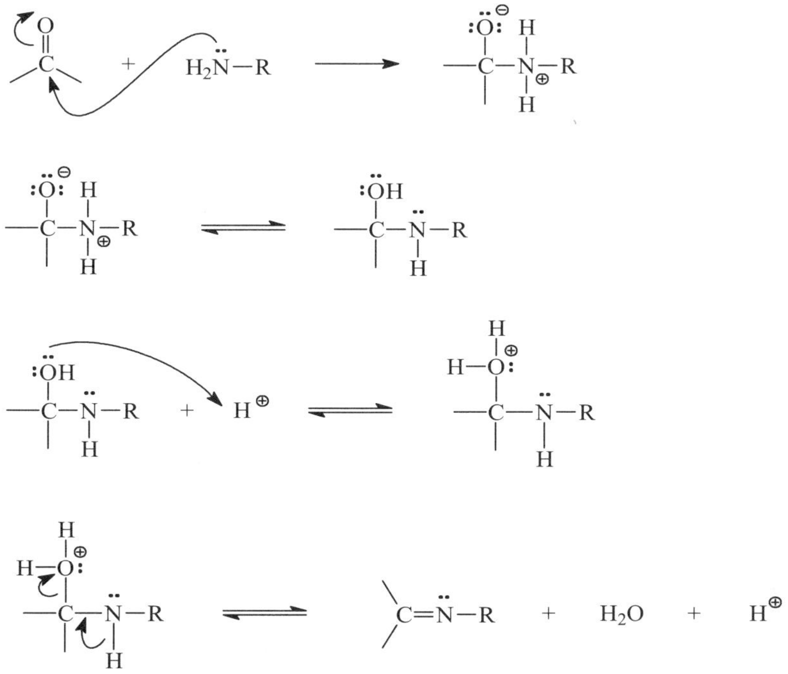

(mechanism shown: carbonyl + H₂N—R → tetrahedral intermediate with O⁻ and NH₂⁺R; proton transfer to give C(OH)(NHR); protonation of OH to give C(OH₂⁺)(NHR); loss of water and proton to give C=N—R + H₂O + H⁺)

☐☐ **When an Aldehyde is reacted with a 2° Amine, an EnAmine is produced. What is the best way to drive this reaction to completion?**

The other product is water, so removing water with a Dean-Stark trap will drive the reaction to completion.

☐☐ **Predict the major product(s) in each of the following reactions.**

a) 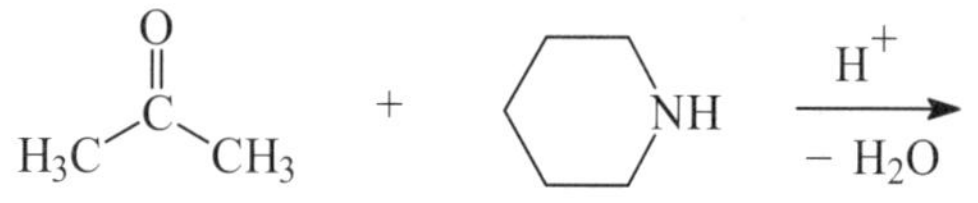 $H_3C-CO-CH_3$ + piperidine (NH) $\xrightarrow{H^+, -H_2O}$

b) 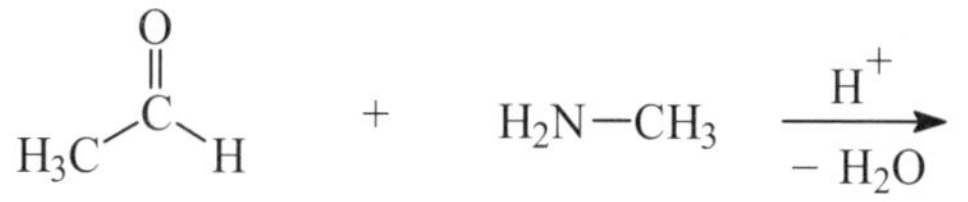 H_3C-CHO + H_2N-CH_3 $\xrightarrow{H^+, -H_2O}$

c) 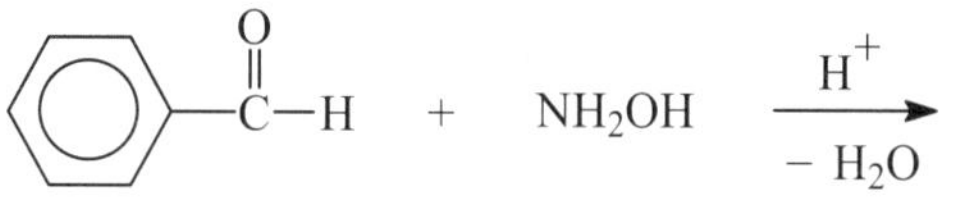 Ph-CHO + NH_2OH $\xrightarrow{H^+, -H_2O}$

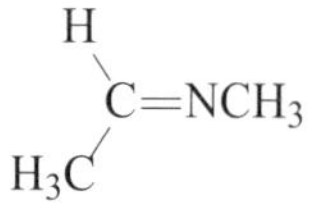

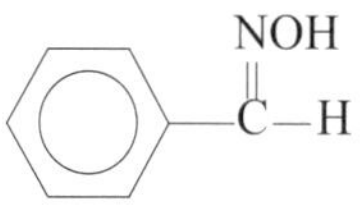

d) 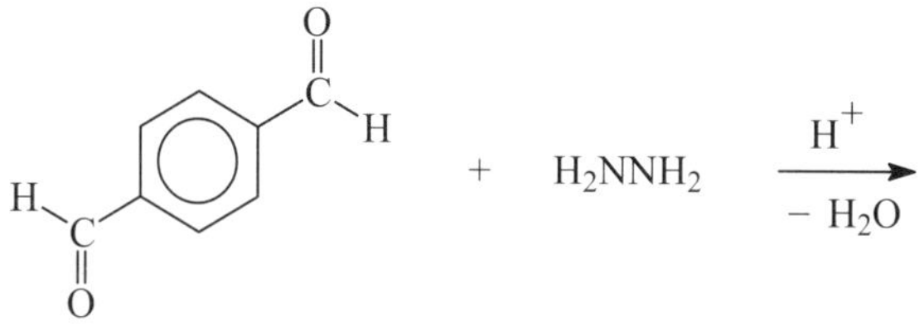 terephthalaldehyde (OHC-C6H4-CHO) + H_2NNH_2 $\xrightarrow{H^+, -H_2O}$

a) 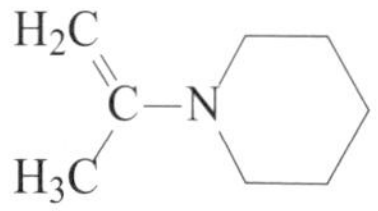 $H_2C=C(CH_3)-N(piperidinyl)$

b) $H_3C-CH=NCH_3$

c) Ph-CH=NOH, a Oxime

d) 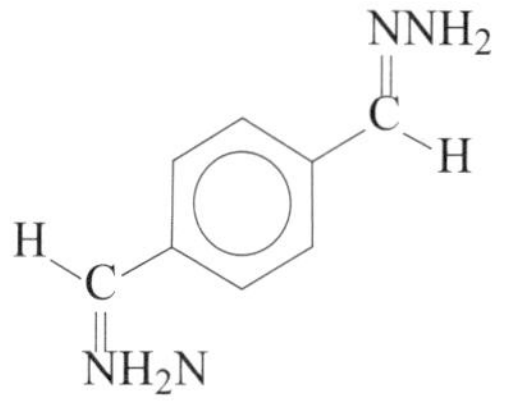 $H_2NN=CH-C_6H_4-CH=NNH_2$

☐☐ **In water over 95% of dissolved Formaldehyde is converted to the hydrate, but under the same conditions, only 1% of Acetone molecules are hydrated. Why is this?**

Hydrate formation is an equilibrium process and is controlled by a combination of electronic and steric factors. Any factors which stabilize the carbonyl compound will decrease the equilibrium constant for hydration, as will any factors that destabilize the hydrate. Alkyl groups are generally electron donating, and are capable of stabilizing the partial positive charge on a carbonyl Carbon. Acetone has two alkyl substituents attached to its carbonyl Carbon while Formaldehyde has none. This suggests that Acetone is lower in energy than Formaldehyde. Electronic factors here therefore favor the hydrate in the case of

Formaldehyde and favor the Ketone in the case of Acetone. If we look at steric factors, we notice that the substituents in a carbonyl compound are 120° apart while they are only 109.5° apart in the hydrate. Thus we would expect smaller substituents, such as Hydrogen, to favor the hydrate, while bulkier substituents, such as alkyl groups, would favor the carbonyl. Formaldehyde's substituents are both Hydrogens, while Acetone has two alkyl groups. Thus steric factors also favor the hydrate in the case of Formaldehyde, and favor the Ketone in the case of acetone.

❏❏ **For each of the following compounds, draw the two possible enol structures. Which enol form will predominate at equilibrium? Why?**

a) [2-methylcyclohexanone]   b) $H_3C-\underset{CH_3}{\underset{|}{C}}\overset{H}{\overset{|}{-}}\overset{O}{\overset{\|}{C}}-CH_2CH_3$ c) [cyclohex-2-enone]

a) [Major: enol with more substituted double bond] [Minor]

The more stable isomer is the one in which the double bond is most highly substituted.

b) $H_3C-\underset{CH_3}{\underset{|}{C}}=\overset{OH}{\overset{|}{C}}-CH_2CH_3$ Major $H_3C-\underset{CH_3}{\underset{|}{C}}\overset{H}{\overset{|}{-}}\overset{OH}{\overset{|}{C}}=CHCH_3$ Minor

Again, the more stable isomer is the one in which the double bond is most highly substituted.

c) [Major: conjugated phenol-like enol] [Minor]

The more stable isomer is the one in which the enolate double bond is conjugated.

❏❏ **In the presence of dilute acid, optically active (R)-4-Ethyl-3-pentanone rapidly loses all optical activity. How is this possible?**

Under acidic conditions Aldehydes and Ketones with an α-Hydrogen are in equilibrium with their enol forms. Because the Chiral Center of 4-Ethyl-3-pentanone is in a position α to the carbonyl, the chirality is destroyed as the Ketone converts to the Achiral Enol form. Reprotonation to regenerate the Ketone can occur from either side of the enol double bond, so the result is racemization of the material.

Chiral $CH_3CH_2\underset{O}{\overset{\|}{C}}-\underset{CH_3}{\overset{CH_2CH_3}{\underset{|}{\overset{|}{C}}}}\cdots H$ $\xrightarrow{H^+}$ $CH_3CH_2\underset{OH}{\underset{|}{C}}=\overset{CH_2CH_3}{\underset{}{C}}CH_3$ Achiral

❏❏ **Chloral (Trichloroaceteldehyde) very readily undergoes hydrolysis in the presence of water to form Chloral Hydrate, a knockout drug otherwise known as a "Mickey Finn." Why is this hydrolysis so easy?**

Reaction to form the hydrate occurs readily due to electronic factors. The trichloromethyl group attached to the carbonyl in chloral acts as a strong electron withdrawing group, destabilizing the partial positive charge on the carbonyl Carbon. This raises the energy of the Aldehyde, favoring the hydrate in the equilibrium reaction.

☐☐ **Ethylene Glycol is often used to form a protecting group for Aldehydes and Ketones. What is the structure of this protecting group, and how does it work?**

The Ethylene Glycol reacts with the carbonyl under acidic conditions to give a 5-membered ring acetal. Like other Ethers, acetals are relatively unreactive under a wide variety of reaction conditions.

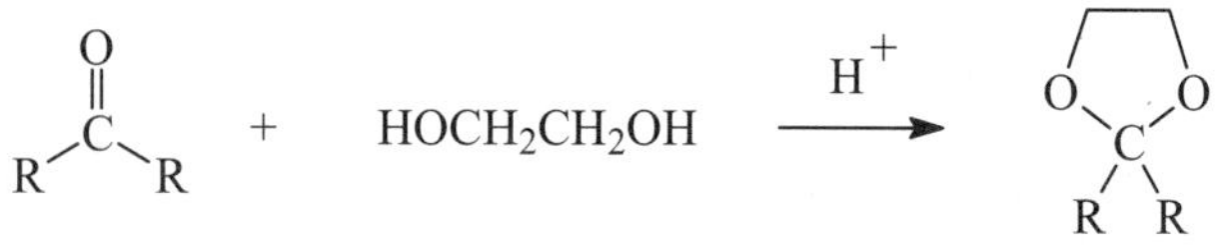

$$R-C(=O)-R + HOCH_2CH_2OH \xrightarrow{H^+} \text{5-membered ring acetal}$$

☐☐ **What is the primary organic product in each of the following reactions?**

a)  $CH_3CH_2-C(=O)H + Cl_2 \xrightarrow{H^+}$

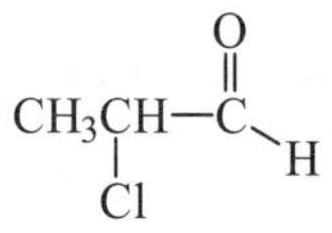

b) 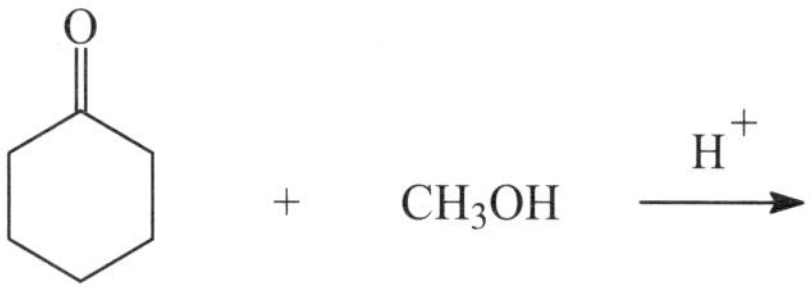 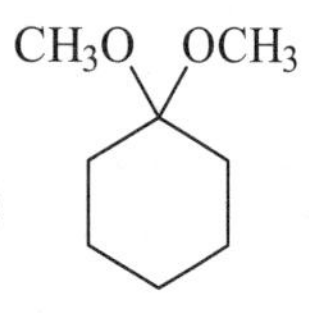 cyclohexanone $+ CH_3OH \xrightarrow{H^+}$

c)  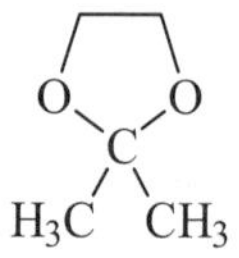 $H_3C-C(=O)-CH_3 + HOCH_2CH_2OH \xrightarrow{H^+}$

d) 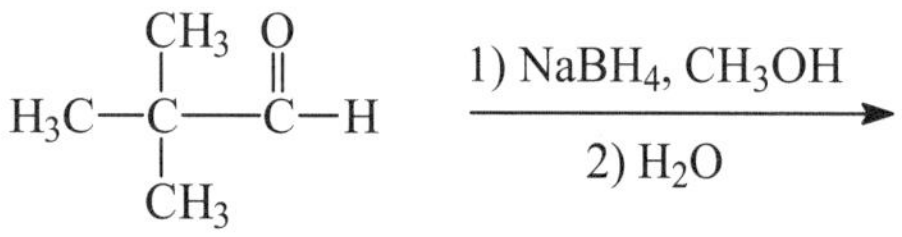 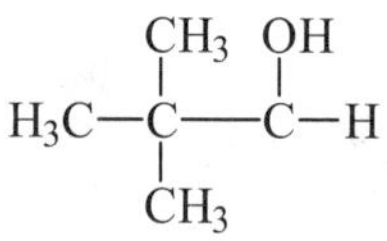 $H_3C-C(CH_3)_2-C(=O)-H \xrightarrow{\text{1) NaBH}_4\text{, CH}_3\text{OH}}_{\text{2) H}_2\text{O}}$

e) 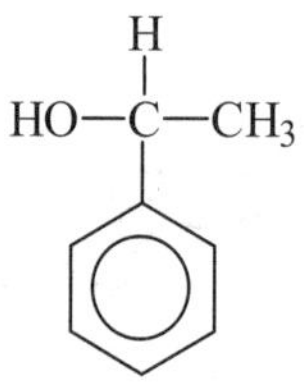 benzaldehyde $\xrightarrow{\text{1) CH}_3\text{MgBr, ether}}_{\text{2) H}_3\text{O}^+}$

a) $CH_3CH(Cl)-C(=O)H$

b) cyclohexane with CH_3O and OCH_3 at same carbon

c) 5-membered acetal ring with H_3C and CH_3 on the acetal carbon

d) $H_3C-C(CH_3)_2-CH(OH)-H$

e) phenyl-CH(OH)-CH_3

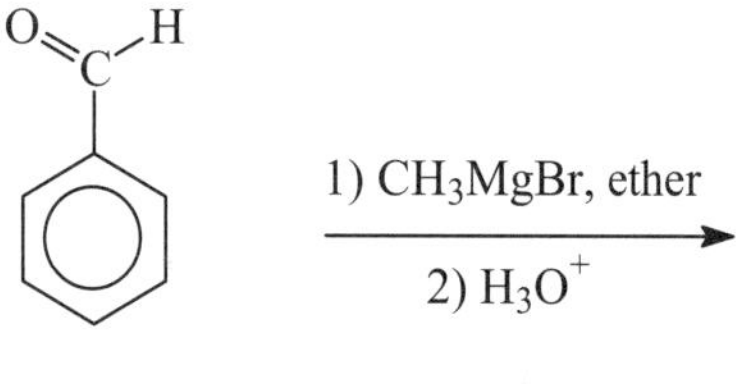

☐☐ What is an Aldol Condensation?

An Aldol Condensation is a nucleophilic addition of an Aldehyde or Ketone enolate to the carbonyl group of another Aldehyde or Ketone molecule. This is followed by dehydration to give and α,β-unsaturated carbonyl compound. The first step is an aldol addition. The second step, which produces water as a by-product, is where the condensation term comes from.

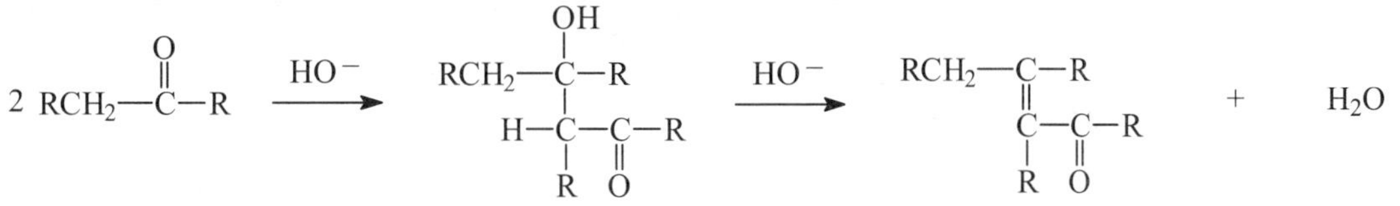

☐☐ What is the mechanism of an Aldol Condensation reaction?

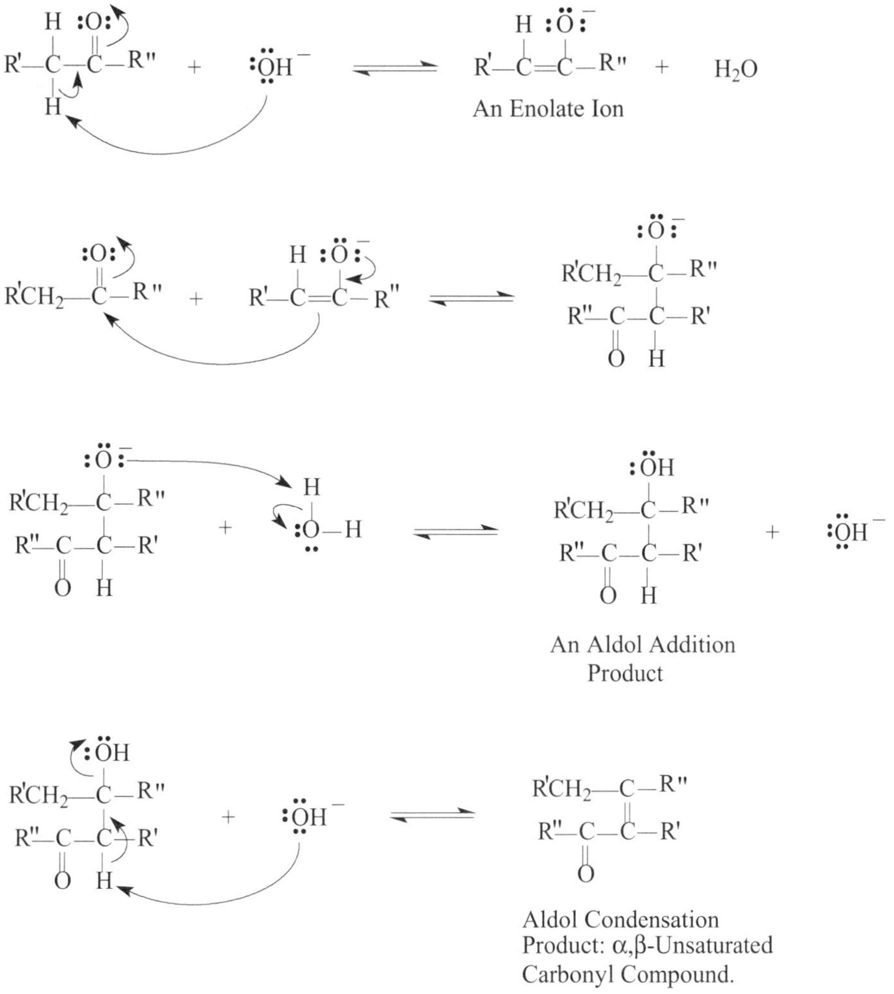

☐☐ Why are mixed Aldols not usually synthetically useful?

Because a mixture of several products usually results. If both compounds are capable of forming enolates, then a total of four products will be formed, two from self-addition, and two from mixed addition.

☐☐ Under what conditions would a mixed Aldol reaction be useful?

If one of the carbonyl compounds could not form an enolate. Formaldehyde and Aryl Aldehydes are often used as one component of such reactions because they have no α protons.

❏❏ **When Ketones are used in an Aldol reaction, the equilibrium of the addition reaction favors the starting Ketone over the initial addition product. Yet aldol reactions of Ketones are very commonly used in synthesis. How is this possible?**

Though the equilibrium of the initial addition usually favors the Ketone starting materials, Aldol condensation reactions of Ketones can be driven to completion if conditions are chosen so as to favor the second dehydration step. The dehydration depletes the initial addition product, thus forcing the equilibrium of the addition step to the right.

❏❏ **What is the Tollen's test?**

It is a qualitative test for Aldehydes. Treatment of an Aldehyde with Tollen's reagent ($Ag(NH_3)_2^+NO_3^-$) gives a Carboxylic Acid and metallic silver. In a glass test vessel, such as a test tube, this silver plates out on the glass as a mirror.

❏❏ **The pK_a of the α Hydrogens of Acetone is 20, but the pK_a of the central α Hydrogens of 2,4-Pentanedione is 10. Why is there such a difference?**

The Hydrogens of 2,4-Pentanedione are easier to remove because the anionic product is more stable. The Enolate Ion of Acetone has two resonance forms, one with the negative charge on a Carbon, and one with the charge on an Oxygen. The enolate ion of 2,4-Pentanedione, on the other hand, has three resonance forms, two of which place the negative charge on an Oxygen. The additional resonance structure provides considerable stabilization to the anion, making it easier to form.

❏❏ **Halogenation of Ketones and Aldehydes at the α position is usually carried out under acidic conditions, even though the reaction would be expected to occur more rapidly under basic conditions. Why is this?**

Under basic conditions the reaction is hard to control. Under acidic conditions monohalogenation occurs. Under basic conditions polyhalogenation is more common, and other side reactions may occur.

❏❏ **What is the Haloform reaction?**

On treatment with Br_2, Cl_2, or I_2 in aqueous base solution, Methyl Ketones undergo a cleavage reaction to produce CHX_3, a haloform. The haloform reaction using Iodine can be used as a test for Methyl Ketones. The formation of a yellow precipitate (Iodoform) is a positive test.

❏❏ **Acetaldehyde, when allowed to stand in deuterated water, yields a product in which all of the protons of the Methyl group are replaced by Deuterium. How is this possible?**

Ketones and Aldehydes in aqueous solution are in a dynamic equilibrium with their enol forms. This means that the alpha protons are constantly being lost and replaced from solution. In deuterated water, the protons of the Acetaldehyde are replaced with the more abundant Deuterium atoms.

❏❏ **What is a Michael addition?**

A Michael addition is the conjugate (1,4) addition of an anion to an α,β-unsaturated carbonyl compound. The anion in this reaction is very often an enolate ion.

❏❏ **Nucleophilic addition to an Alkene is usually rare. Why does it occur in α,β-unsaturated carbonyl compounds?**

Addition of a nucleophile to an ordinary Alkene produces an unstable carbanion. When addition occurs at the double bond of an α,β-unsaturated carbonyl compound, the product is an enolate, which is considerably more stable.

❏❏ **What other terms are used to describe a Michael addition?**

It is also known as a Michael reaction, and falls into the category of conjugate addition, which is also known as 1,4 addition.

❏❏ **What types of nucleophiles favor 1,4 addition?**

1,4 Addition is usually observed with weakly basic nucleophiles.

❏❏ **Conjugate addition is said to be thermodynamically controlled while 1,2 addition is kinetically controlled. What does this mean?**

Under conditions in which there is no equilibrium between the products (irreversible reactions), the one which is produced at the fastest rate is the one that will be obtained in the highest yield. This is known as kinetic control. Generally, the 1, 2 addition reaction is faster than 1,4 addition, so 1,2 addition is the product under conditions of kinetic control. If an equilibrium is established between products, then the rate at which they are intitially produced doesn't matter. The final product yield will be based upon the relative stabilities of the products; the lower energy product will be the one obtained in the highest yield. This is known as thermodynamic control. The 1,4, or conjugate addition product is more thermodynamically stable, so it is the product under conditions of thermodynamic control. Strongly basic nucleophiles tend to result in 1,2 addition because they are unlikely to come off once they are attached (they are poor leaving groups), while weakly basic nucleophiles favor 1,4 addition because they go on and off the carbonyl compound readily (they are good leaving groups) and result in equilibrium conditions.

❏❏ **Why are Lithium Diorganocuprate reagents unique nucleophiles for reaction with Ketones or Aldehydes?**

They tend to undergo 1,4 addition with α,β-unsaturated carbonyl compounds, unlike Grignards, which attack the carbonyl directly.

❏❏ **What would be the structure of each of the following derivatives?**

a) Benzaldehyde oxime   b) Acetaldehyde 2,4-dinitrophenylhydrazone   c) Acetone semicarbazone

❏❏ **Nucleophilic addition of Methyl Magnesium Bromide to 3-Hexanone gives a product with a Chiral Center. Would you expect the resulting product to be optically active?**

No, because the Methyl group can be added to either face of the carbonyl with equal probability. The result will be a Racemic mixture, and therefore optically inactive.

❏❏ **If Methyl Magnesium Bromide reacted with the molecule shown below, what would be the stereochemistry of the major isolated product? Why?**

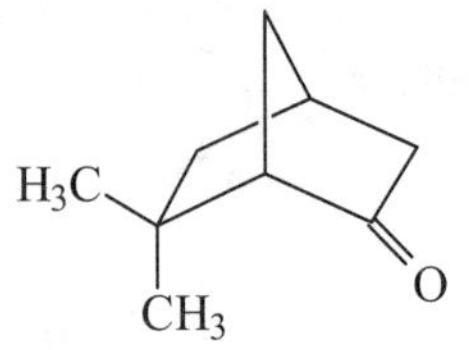

The nucleophile could attack from either the top or the bottom face of the carbonyl. Since the approach from the top face is less hindered, it will probably produce the major product.

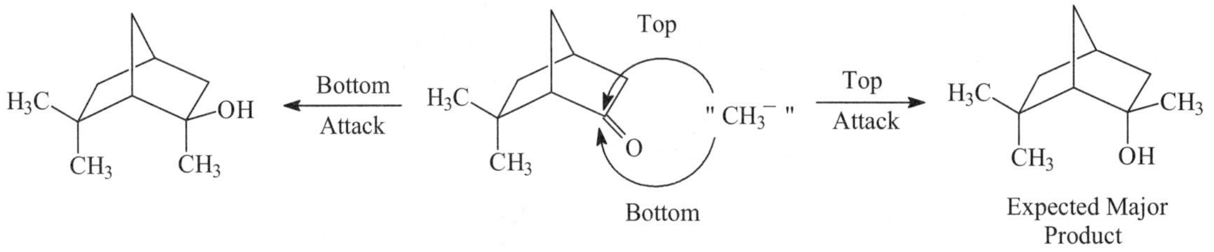

Expected Major Product

❏❏ **Michael addition followed by an Aldol condensation to form a ring is a very useful approach to the synthesis of bicyclic ring systems such as the one shown below. By what name is this procedure known?**

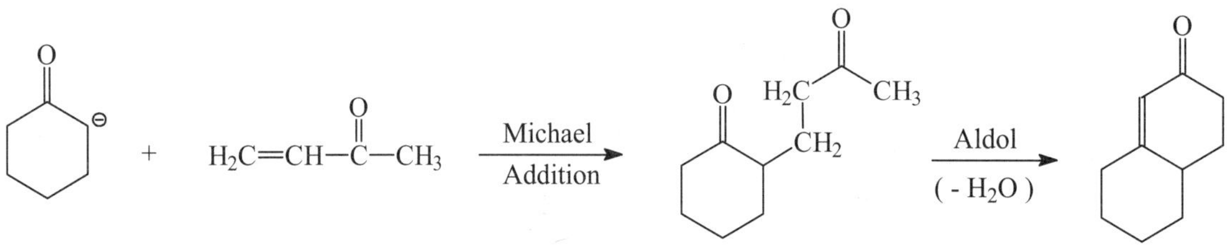

This is an example of a Robinson Annulation reaction.

❏❏ **What size rings can be created by this procedure?**

Five and six membered rings.

❏❏ **What is a Hemiacetal?**

A Hemiacetal is a molecule containing an -OH and an -OR group bonded to the same Carbon. They are formed when one molecule of an Alcohol adds to the carbonyl of an Aldehyde or Ketone. This is the first step in the formation of an acetal, which contains two -OR groups bonded to the former carbonyl Carbon. Hemiacetals are usually formed in only small equilibrium concentrations during acetal formation.

❏❏ **Under what conditions are Hemiacetals the predominant form of a molecule?**

Only when the -OH group and the carbonyl are in the same molecule, and the formation of the Hemiacetal gives a five or six membered ring.

Major isomer present

☐☐ **Match the following molecules with the phrase that best describes them.**

1) 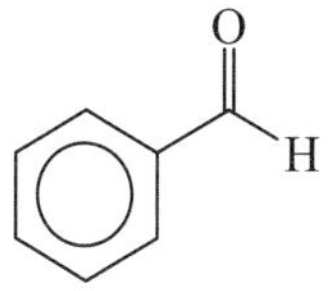 benzaldehyde   a) Possible product of a haloform Rx

2) 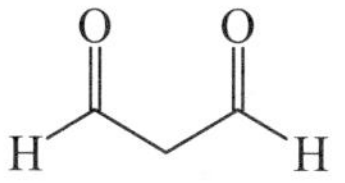 malondialdehyde (OHC-CH₂-CHO)   b) Subject to conjugate addition

3) 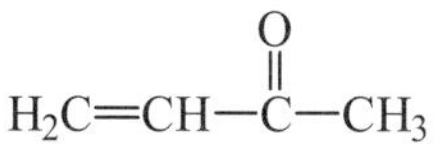 H₂C=CH–C(=O)–CH₃   c) Good for crossed aldols

4) 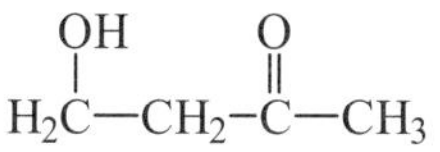 H₂C(OH)–CH₂–C(=O)–CH₃   d) Slightly enhanced acidity

5) 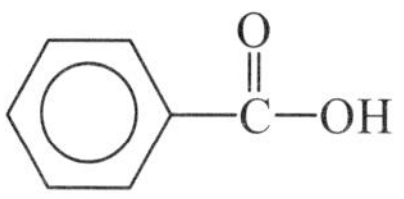 benzoic acid   e) Product of an aldol addition

1c, 2d, 3b, 4e, 5a.

☐☐ **What would each of the following reagents do to a Ketone?**

a) Zn (Hg) / HCl   **b) H₂NNH₂**   **c) NaBH₄**   **d) CrO₃**

(a) The carbonyl of the Ketone would be reduced to a methylene group, (b) This reagent would convert the Ketone to a hydrazone derivative, (c) Sodium borohydride would reduce the Ketone to a secondary Alcohol, d) There would be no reaction. Ketones are resistant to oxidation except under extreme conditions.

☐☐ **What is the Wolff-Kishner reduction?**

It is a method for converting a carbonyl to a methylene group. It involves reacting the Aldehyde or Ketone with Hydrazine and concentrated KOH under refluxing conditions in a high boiling solvent such as Diethylene Glycol.

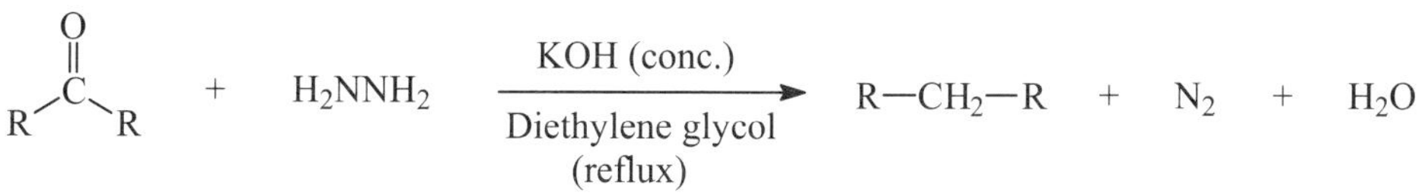

R–C(=O)–R + H₂NNH₂ →[KOH (conc.) / Diethylene glycol (reflux)] R–CH₂–R + N₂ + H₂O

☐☐ **What is a Baeyer-Villiger oxidation?**

The Baeyer-Villiger oxidation is a method for oxidizing Ketones to Esters using peroxyacids. The reaction in effect inserts an Oxygen into the bond between the carbonyl and the α Carbon.

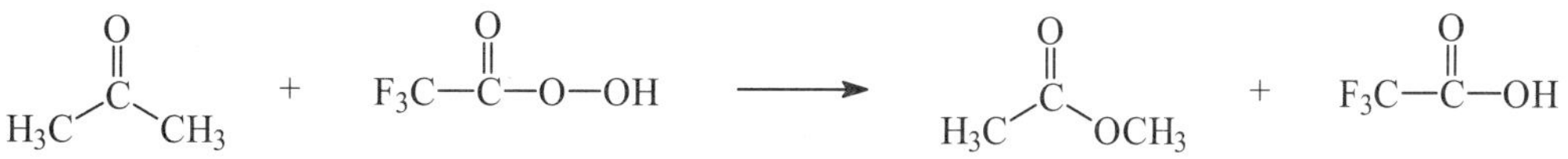

H₃C–C(=O)–CH₃ + F₃C–C(=O)–O–OH ⟶ H₃C–C(=O)–OCH₃ + F₃C–C(=O)–OH

☐☐ **What is the mechanism of the Baeyer-Villiger oxidation?**

[Mechanism scheme showing the three steps of the Baeyer-Villiger oxidation, with a ketone reacting with a peroxyacid to form a tetrahedral intermediate, proton transfer, then migration to give an ester and a carboxylic acid.]

☐☐ **When an Aldehyde or Ketone is refluxed with amalgamated Zinc in concentrated HCl, the carbonyl is reduced to a methylene group. By what name is this reaction known?**

This is known as a Clemmensen reduction.

☐☐ **The Wolff-Kishner reduction and the Clemmensen reduction compliment each other very well. Why is this?**

The Clemmensen reduction is carried out in concentrated acid while the Wolff-Kishner reduction is carried out in concentrated base. If a molecule cannot be reduced using the Clemmensen reaction due to sensitivity to acid, then it can usually be reduced by the Wolff-Kishner method. The reverse also applies; molecules sensitive to base, for which the Wolff-Kishner reaction is inappropriate, may be reduced by the Clemmensen reaction. Between the two approaches, the carbonyls of most Ketones and Aldehydes may be reduced to methylene groups.

☐☐ **Since Enolate ions are nucleophilic, one would expect them to react with Haloalkanes to give alkylation products. Why is this reaction difficult to use with normal Aldehydes and Ketones?**

With typical Aldehydes and Ketones and typical bases, Aldol condensation competes with alkylation. If alkylation does successfully compete, then it is difficult to limit the reaction to a single alkylation. In short, the reaction either fails completely or gives a mixture of products.

☐☐ **Under what conditions can the alkylation of enolates be carried out succesfully?**

The alkylation works well when β-diketones are used as precursors of the enolate. β-Diketones are acidic enough that they can be quantitatively deprotonated with relatively weak bases, and then alkylated with methyl or primary Alkyl Halides. No significant self-condensation occurs under these conditions. If normal Ketones or Aldehydes are to be used, they must be converted completely to their enolate forms by using very strong bases such as Sodium Amide or Potassium Hydride. Protic solvents are usually avoided in these reactions because they would either react with the base or reprotonate the enolate. Aprotic solvents such as THF, DMF and liquid Ammonia are typically used.

What would be the primary organic product in each of the following reactions?

a) 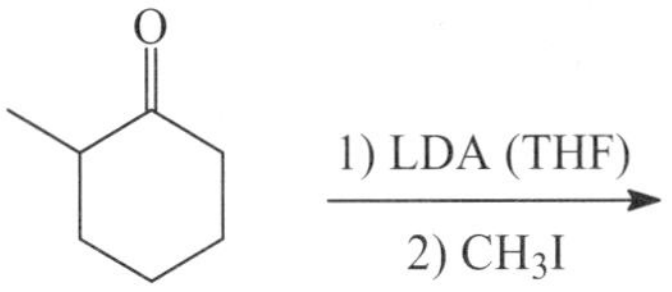 [2-methylcyclohexanone] $\xrightarrow{\text{1) LDA (THF)} \\ \text{2) CH}_3\text{I}}$

b) 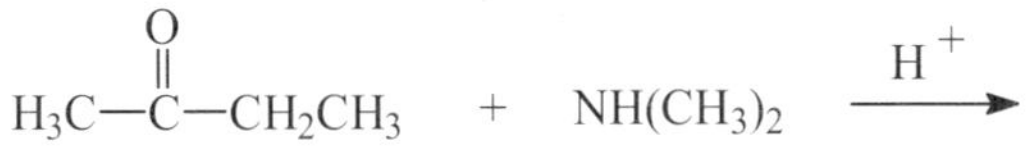 $H_3C-\overset{O}{\underset{||}{C}}-CH_2CH_3$ + $NH(CH_3)_2$ $\xrightarrow{H^+}$

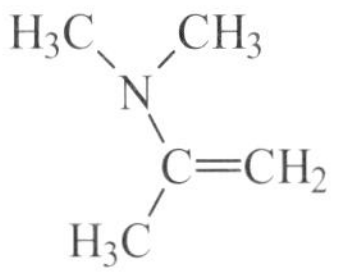

c) 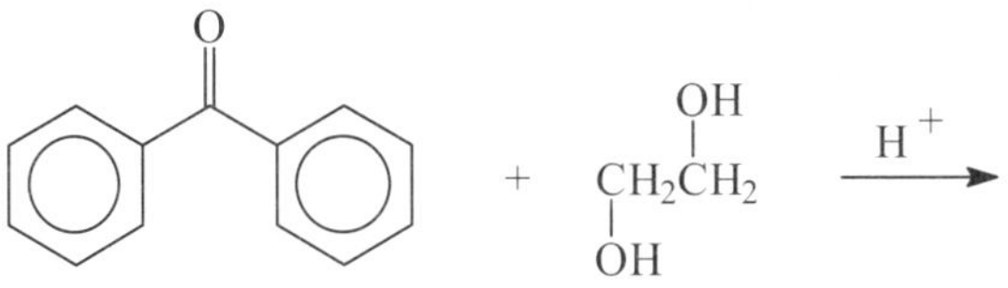 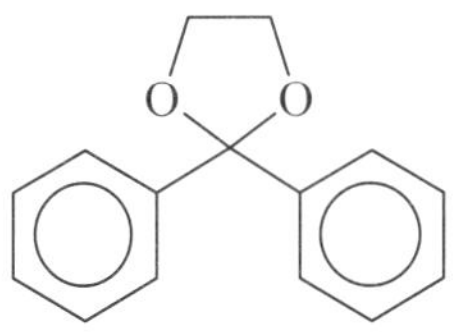 [benzophenone] + $\underset{\underset{OH}{|}}{\overset{\overset{OH}{|}}{CH_2CH_2}}$ $\xrightarrow{H^+}$

d)  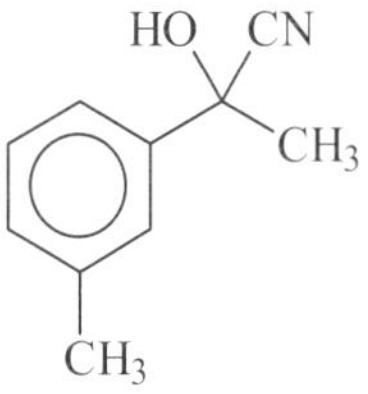 [3'-methylacetophenone] + HCN $\xrightarrow{\text{NaCN, pH 10} \\ H_2O}$

a) 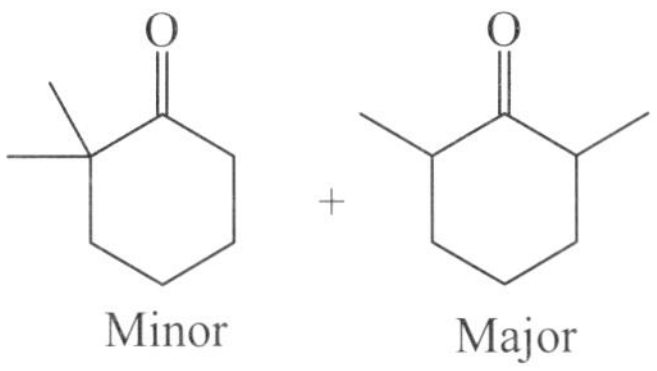 [2,2-dimethylcyclohexanone] Minor + [2,6-dimethylcyclohexanone] Major

b) $\underset{H_3C}{\overset{H_3C}{\diagdown}}\overset{CH_3}{\underset{|}{N}}\\ C=CH_2$

c) [2,2-diphenyl-1,3-dioxolane]

d) [3-methylphenyl cyanohydrin with CH$_3$, HO, CN]

What product is obtained when a Thiol is reacted with an Aldehyde or Ketone in the presence of acid?

Thiols react very much like Alcohols in this case. The products are Thioacetals.

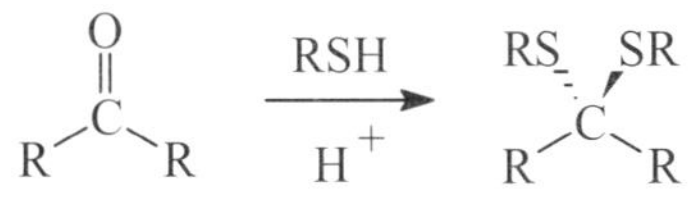

$\underset{R}{\overset{O}{\underset{||}{R-C-R}}} \xrightarrow[H^+]{RSH} \underset{R}{\overset{RSSR}{\underset{|}{C}}}R$

❏❏ **The most common Thiol for preparing Thioacetals is 1,3-Propanedithiol. For what are the 1,3-Dithiane products of this reaction used?**

Dithianes prepared with Aldehydes are slightly acidic (pK$_a$ of about 31) and can be deprotonated with butyl lithium. The anions produced make very good nucleophiles and will react with primary Alkyl Halides, Aldehydes, and Ketones. The Propanethiol protecting group can then be removed with Mercuric Chloride (HgCl$_2$) in aqueous Acetonitrile to regenerate the original carbonyl. Thus the process provides a method of converting Aldehydes to Ketones.

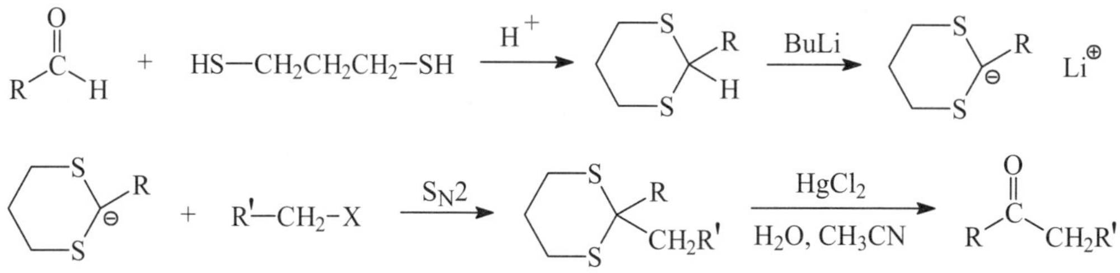

❏❏ **The reaction between Triphenylphosphine and Epoxides is said to be similar to the Wittig reaction. What do they have in common?**

Both reactions go through a betaine/oxaphosphetane intermediate and yield an Alkene as the product.

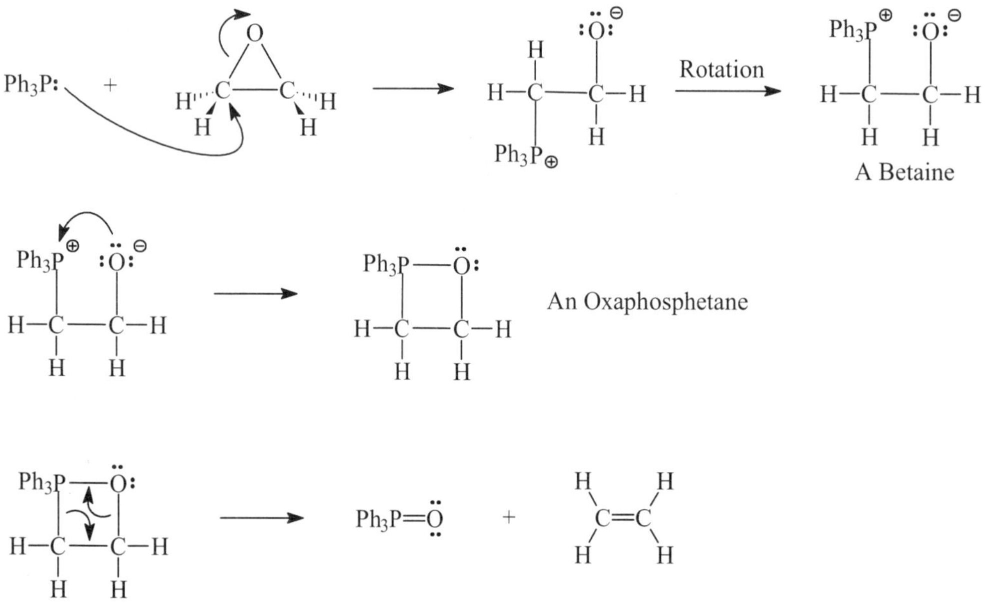

❏❏ **Acetone reacts with Halogens in the presence of acid to give products halogenated in the α position. What are the relative rates for Chlorine, Bromine, and Iodine?**

The Halogens all react at exactly the same rate.

❏❏ **What do these relative rates of Halogenation tell you about the mechanism of this reaction?**

The Halogen does not take part in the reaction until after the rate determining step.

☐☐ **What is the mechanism for the α halogenation of Acetone in the presence of acid?**

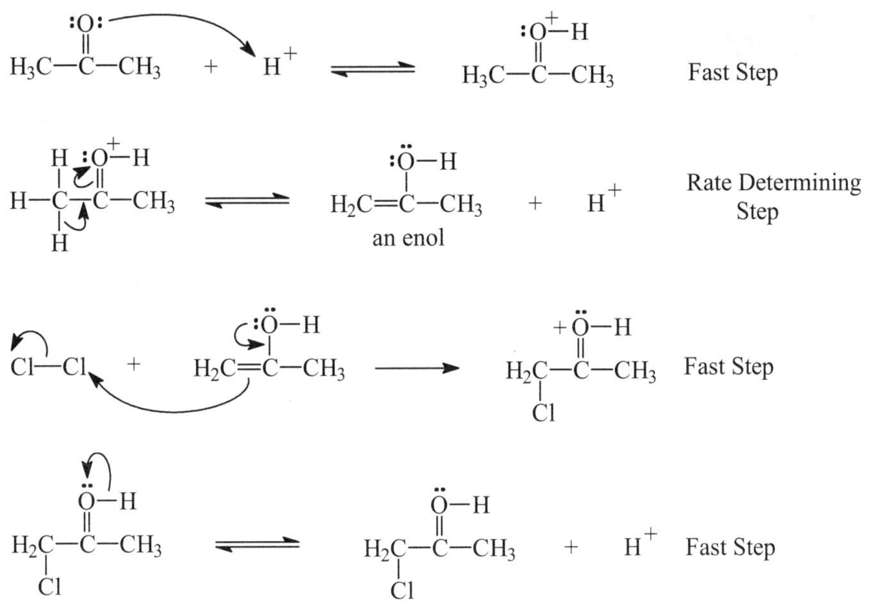

$$H_3C-\underset{\substack{\|\\ :O:}}{C}-CH_3 + H^+ \rightleftharpoons H_3C-\underset{\substack{\|\\ :\overset{+}{O}-H}}{C}-CH_3 \quad \text{Fast Step}$$

$$H-\underset{\substack{|\\H}}{\overset{H}{C}}-\underset{\substack{\|\\ :\overset{+}{O}-H}}{C}-CH_3 \rightleftharpoons H_2C=\underset{\substack{|\\ :\ddot{O}-H}}{C}-CH_3 + H^+ \quad \text{Rate Determining Step}$$

an enol

$$Cl-Cl + H_2C=\underset{\substack{|\\ :\ddot{O}-H}}{C}-CH_3 \longrightarrow H_2C-\underset{\substack{|\\ Cl}}{\overset{\substack{+\ddot{O}-H\\\|}}{C}}-CH_3 \quad \text{Fast Step}$$

$$H_2C-\underset{\substack{|\\ Cl}}{\overset{\substack{\ddot{O}-H\\\|}}{C}}-CH_3 \rightleftharpoons H_2C-\underset{\substack{|\\ Cl}}{\overset{\substack{\ddot{O}-H\\|}}{C}}-CH_3 + H^+ \quad \text{Fast Step}$$

CARBOXYLIC ACID AND CARBOXYLIC ACID DERIVATIVE PEARLS

It is disconcerting to reflect on the number of students we have flunked in Chemistry for not knowing what we later found to be untrue.
Quoted in Robert L. Weber, Science With A Smile

❏❏ **What is a Carboxylic Acid?**

Any compound that contains a carboxyl group:

$$-C(=O)OH$$

a carboxyl group

❏❏ **What classes of compounds are considered to be derivatives of Carboxylic Acids?**

The derivatives of Carboxylic Acids include Acid Halides, Esters (including Lactones), Acid Anhydrides, Amides (including Lactams), and Nitriles.

❏❏ **How are Carboxylic Acids named according to IUPAC standard nomenclature?**

For open chain Carboxylic Acids, the longest chain containing the carboxyl group is chosen as the parent compound. The -e ending of the corresponding Alkane is replaced with the suffix -oic acid. Substituents are named appropriately and their position indicated by giving the number of the Carbon to which they are attached. The Carbon of the carboxyl group is numbered as Carbon-1. Compounds in which the carboxyl group is bonded to a ring are named by giving the name of the ring and adding the suffix -carboxylic acid. The ring Carbon bearing the carboxyl group is considered to be Carbon-1 for numbering purposes.

❏❏ **What would be the IUPAC standard name of each of the following compounds?**

a) $CH_3CH_2CH(CH_3)-C(=O)-OH$

b) $HO-C(=O)-CH_2CH_2CH_2-C(=O)-OH$

c) $H_3C-C(=O)-O-CH_2CH_3$

d) $H_3C-C(=O)-O-C(=O)-CH_3$

e) $CH_3CH(CH_3)-C(=O)-Cl$

f) $H_3C-C(=O)-O-C(=O)-CH_2CH_2CH_3$

(a) 2-Methylbutanoic acid, (b) Pentanedioic acid, (c) Ethyl Ethanoate (Ethyl Acetate), (d) Acetic Anhydride, (e) 2-Methylpropanoyl Chloride, (f) Acetic Butanoic Anhydride

❏❏ **What are the most common ways of preparing Carboxylic Acids in the lab?**

Carboxylic acids may be prepared by oxidation of 1° and 2° alkylbenzenes with $KMnO_4$ or Sodium Dichromate, by oxidation of 1° Alcohols and Aldehydes with Jones' reagent, by hydrolysis of Nitriles made by Nucleophilic Substitution of Alkyl Halides with Sodium Cyanide, or by carboxylation of Grignard reagents with CO_2.

❑❑ **What is a Carboxylic Acid Anydride?**

A Carboxylic Acid Anhydride is a molecule in which two acyl groups are bonded to the same Oxygen. Formally, acid anhydrides are made from two molecules of a Carboxylic Acid by removal of a molecule of water.

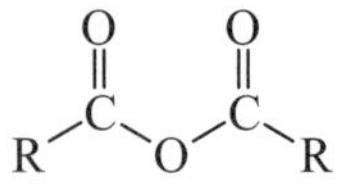

$$R-\overset{O}{\underset{}{C}}-O-\overset{O}{\underset{}{C}}-R$$

A Carboxylic Acid Anhydride

❑❑ **How are acid anhydrides named using IUPAC nomenclature?**

Symmetric anhydrides are named by replacing the word "acid" in the name of the parent Carboxylic Acid with the word "anhydride." Mixed (unsymmetric) anhydrides are named by giving the name of each constituent parent Carboxylic Acid in turn, minus the "acid" ending, and following that with the word "anhydride".

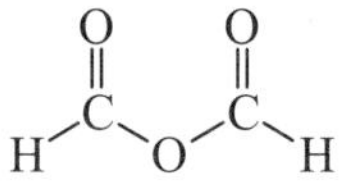

Formic Anhydride 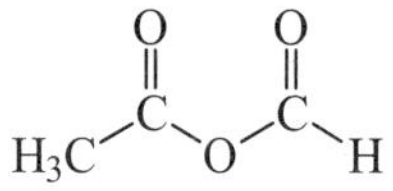   Acetic Formic Anhydride

❑❑ **What is the general structure of an Ester?**

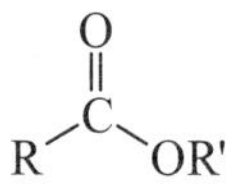

$$R-\overset{O}{\underset{}{C}}-OR'$$

❑❑ **How are Esters named using IUPAC nomenclature?**

The alkyl group attached to the Oxygen is named first, followed by the name of the parent acid with the -ic acid suffix replaced by the ending -ate.

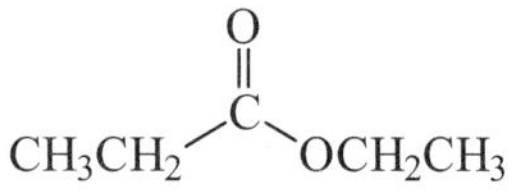

$$CH_3CH_2-\overset{O}{\underset{}{C}}-OCH_2CH_3$$

Ethyl Propanoate

❑❑ **What characteristic physical property do most simple Esters share?**

Many simple Esters are liquids with very pleasant odors. Esters are responsible for the odors associated with a wide variety of flowers and fruits.

❑❑ **How are Amides named in IUPAC nomenclature?**

The -ic acid ending of the parent Carboxylic Acid is replace by the suffix -amide. Any alkyl substituents on the Nitrogen are named and their position is indicated by the prefix N-.

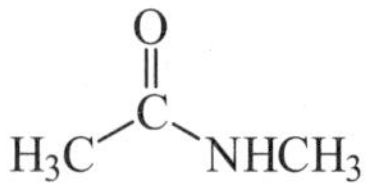

$$H_3C-\overset{O}{\underset{}{C}}-NHCH_3$$

N-Methylacetamide

❑❑ **What is the physical state of most Amides?**

With the exception of some of the Formamides, all Amides are solids at room temperature.

❏❏ **What class of compound would each of the following fall into?**

a) 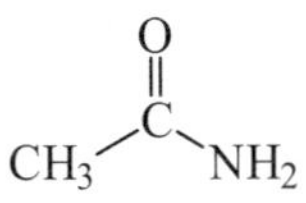 CH₃—C(=O)—NH₂   b) CH₃CH₂—C(=O)—O—C(=O)—CH₂CH₃   c) cyclohexyl—C(=O)—OH

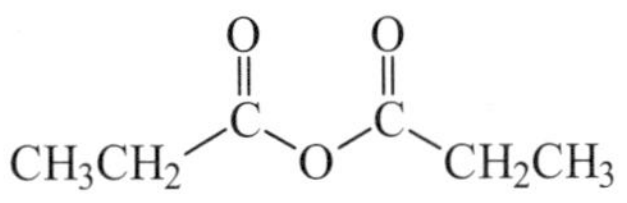

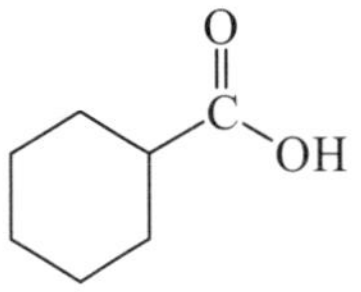

d) 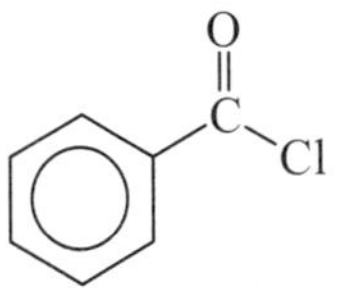 phenyl—C(=O)—Cl   e) 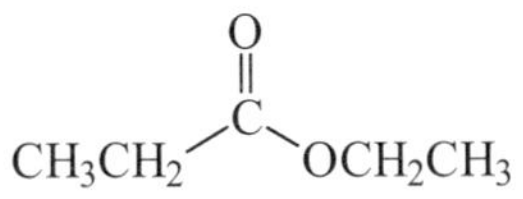 CH₃CH₂—C(=O)—OCH₂CH₃   f) 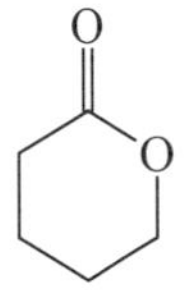 (6-membered lactone ring)

(a) Amide, (b) Acid Anhydride, (c) Carboxylic Acid, (d) Acid Chloride, (e) Ester, (f) Lactone (a cyclic Ester)

❏❏ **Why do Carboxylic Acids have such low pK_a values?**

The anion that results is resonance stabilized and therefore easier to form. There is also an inductive stabilization effect from the partial positive charge on the carbonyl.

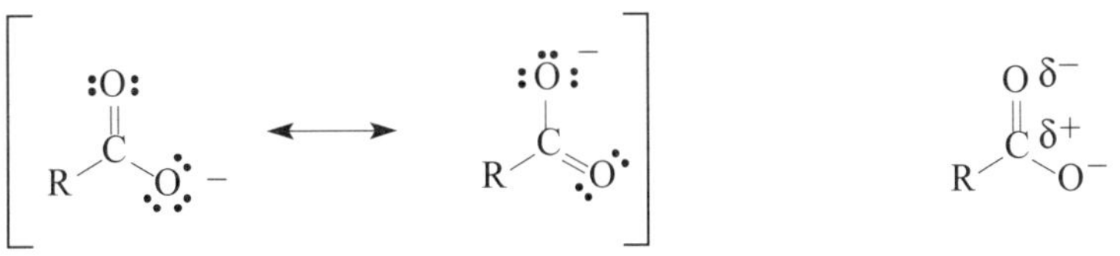

❏❏ **What would be the formula for the acidity constant of acetic acid?**

$$K_a = \frac{[CH_3COO^-][H_3O^+]}{[CH_3COOH]}$$

❏❏ **What is the value of the acidity constant of a typical Carboxylic Acid?**

Carboxylic acids usually have acidity constants, K_a, of around 10^{-5}.

❏❏ **The boiling points of Carboxylic Acids are higher than those of comparable sized hydrocarbons or Oxygen containing organic molecules. Why is this?**

Carboxylic acids are capable of hydrogen bonding, and they do so in a unique way. They form dimers, whose effective size is twice that of a single Carboxylic Acid molecule. This contributes to the increased boiling point.

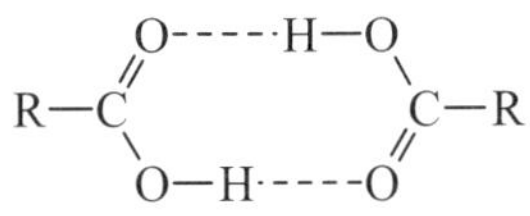

❏❏ **What are the structures of the following molecules?**

a) Propenoic Acid   b) Isopropyl Benzoate

c) Benzonitrile   d) N-Methylacetamide

a) 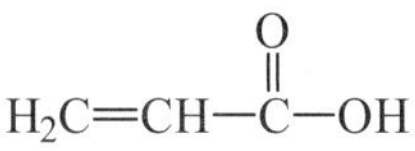 $H_2C=CH-C(=O)-OH$

b) Ph–C(=O)–O–CH(CH$_3$)$_2$

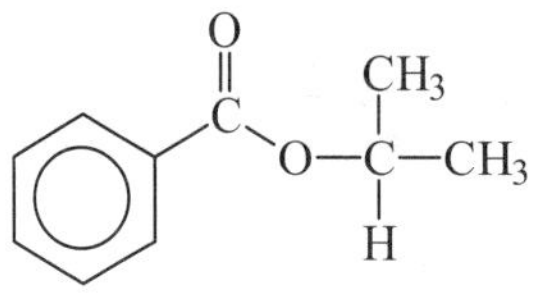

c) Ph–C≡N

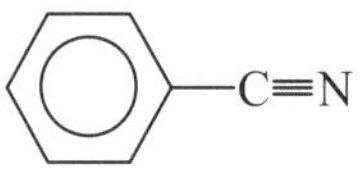

d) 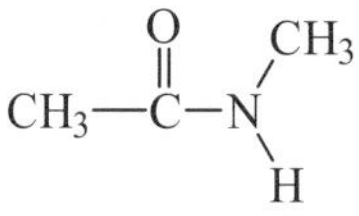 $CH_3-C(=O)-N(CH_3)H$

❏❏ **What are the common names of the compounds shown below?**

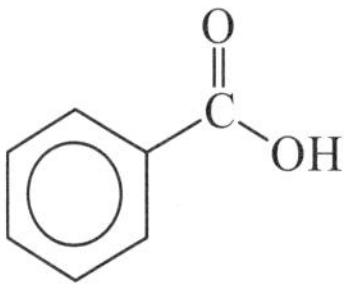

a) 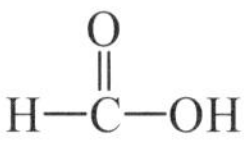 $H-C(=O)-OH$

b) Ph–C(=O)–OH

c) $H_2C=CH-C(=O)-OH$

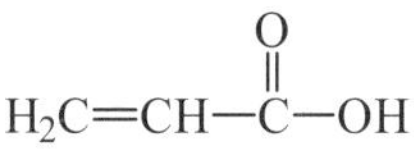

d) 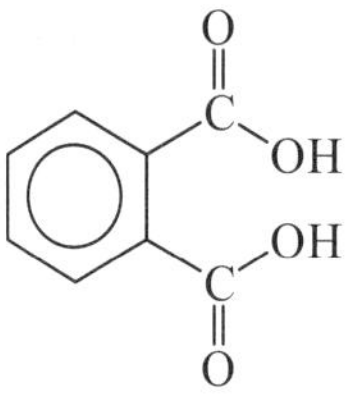 benzene-1,2-dicarboxylic acid (ortho di-COOH on benzene)

(a) Formic Acid, (b) Benzoic Acid, (d) Acrylic Acid, (e) Phthalic Acid

❏❏ **What effect do substituents have on the acidity of Carboxylic Acids?**

Substituents can have very strong effects on the acidity of Carboxylic Acids. Since the dissociation of a Carboxylic Acid is an equilibrium process, any factor which effects the stability the carboxylate anion will shift the equilibrium. Electron withdrawing groups attached to the carboxyl will withdraw electron density by induction and stabilize the carboxylate anion by delocalizing the charge. The Carboxylic Acid will become more acidic. Electron donating substituents attached to the carboxyl group will have the opposite effect, and will result in a less acidic Carboxylic Acid.

❏❏ **Where must an electron withdrawing substituents be placed to have the maximum effect on acidity?**

Since the inductive effect depend strongly on distance, the electron withdrawing substituents must be placed as close to the carboxyl group as possible, preferably in the 2 position.

☐☐ **How would you rank the following Carboxylic Acids in order of increasing acidity?**

a) 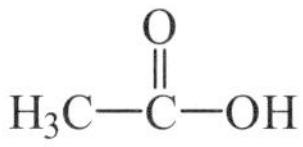 H₃C—C(=O)—OH

b) Cl₃C—C(=O)—OH

c) CH₃CH(CH₃)—C(=O)—OH

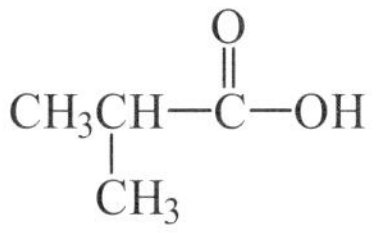

d) 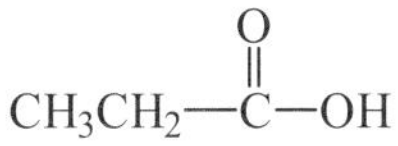 CH₃CH₂—C(=O)—OH

From weakest to strongest acid: c, d, a, b.

☐☐ **The first ionization constant of Oxalic Acid is 6.5 x 10^{-2}. The second is 5.3 x 10^{-5}. Why is there such a difference?**

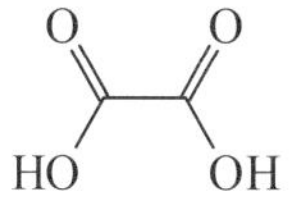

Oxalic Acid: HO—C(=O)—C(=O)—OH

If we consider the first proton, there are several factors that favor its loss. For one, there are two possible sites for deprotonation, statistically increasing the likelihood of proton loss. In addition, the second carbonyl group acts as an electron withdrawing group by induction, thus stabilizing the anion the would be formed on loss of the first proton. The second proton, on the other hand, must be lost from a species that already bears a negative charge. Having two negative charges in such close proximity is not energetically, so the dianion is more difficult to form, and therefore the second proton is not lost as easily as the first.

☐☐ **How would you rank the following substituted benzoic acids in order of increasing acidity?**

a) 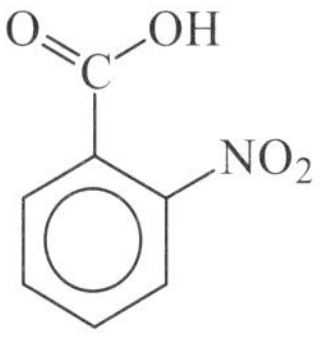 benzoic acid with ortho-NO₂

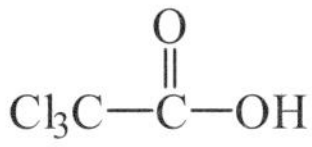

b) 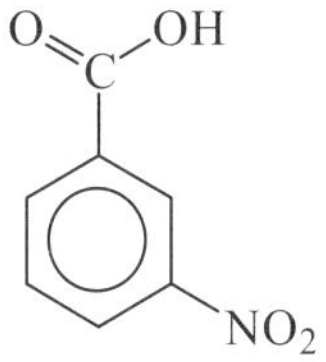 benzoic acid with meta-NO₂

c) 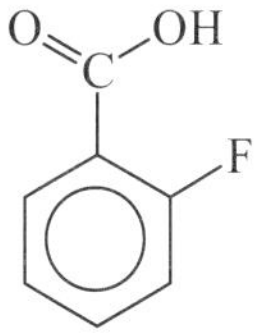 benzoic acid with ortho-F

d) 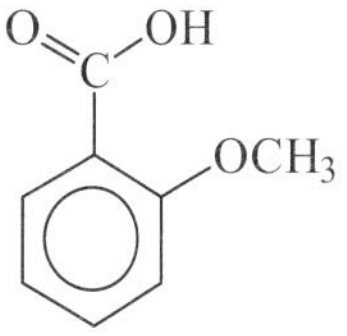 benzoic acid with ortho-OCH₃

a) From strongest to weakest acid: a, d, b, c.

☐☐ **Salts of long chain Carboxylic Acids (C_{12} - C_{18}) are used as soaps. How does this work?**

Long chain Carboxylic Acid salts have a hydrophobic (lipophilic) end (the hydrocarbon chain) and a hydrophilic end (the acid anion.) The hydrophobic ends tend to clump together in water solution, while the hydorphilic ends face out. These structures are known as micelles. Oily materials that normally are water insoluble dissolve in the hydrophobic inner portions of the micelles and may then be washed away.

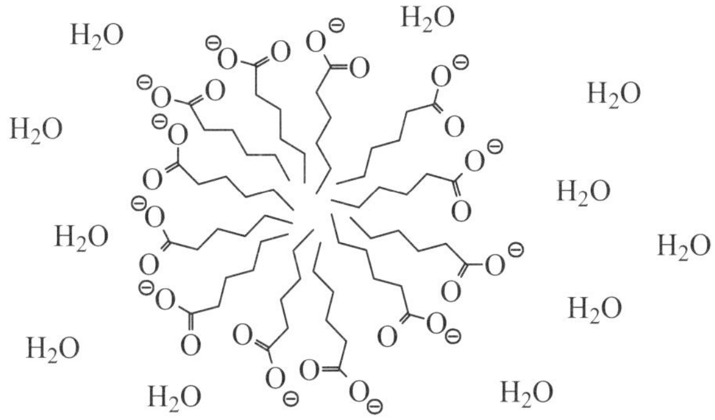

❏❏ **What would be the expected primary organic product(s) in each of the following reactions?**

a) 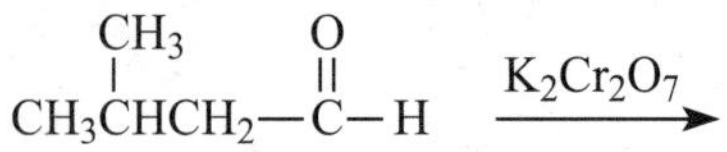 $CH_3CH(CH_3)CH_2-C(=O)-H \xrightarrow{K_2Cr_2O_7}$

b) 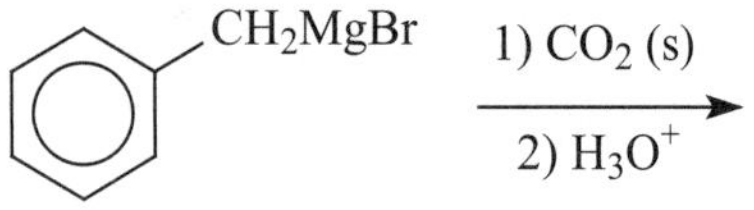 PhCH$_2$MgBr $\xrightarrow{\text{1) } CO_2(s) \quad \text{2) } H_3O^+}$

c) $Br-CH_2CH_2CH_2CH_2-Br \xrightarrow{\text{1) NaCN, H}_2\text{O} \quad \text{2) H}_3\text{O}^+, \Delta}$

d) 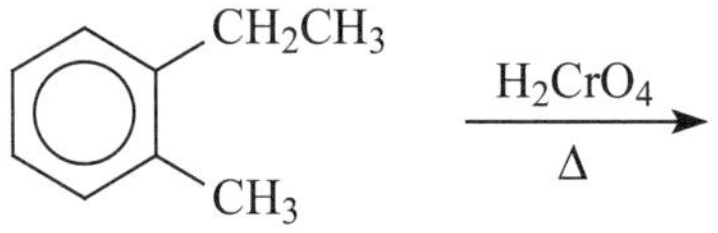 o-ethyltoluene (benzene with CH$_2$CH$_3$ and CH$_3$ substituents) $\xrightarrow{H_2CrO_4, \Delta}$

a) 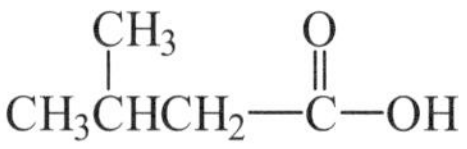 $CH_3CH(CH_3)CH_2-C(=O)-OH$

b) 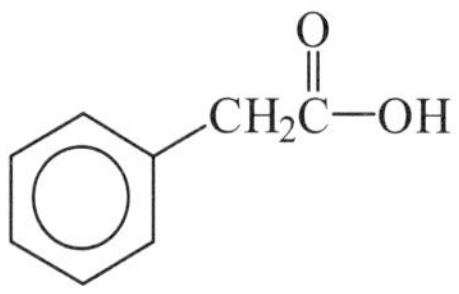 PhCH$_2$C(=O)-OH

c) $HO-C(=O)-CH_2CH_2CH_2CH_2-C(=O)-OH$

d) 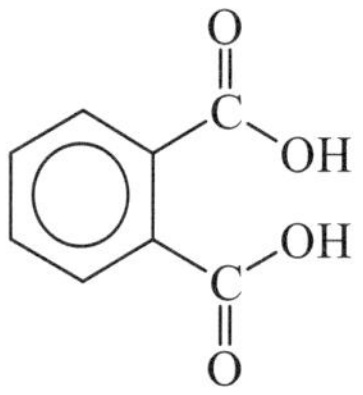 phthalic acid (benzene-1,2-dicarboxylic acid)

❏❏ **In an isotopic labeling experiment, Benzoic Acid was reacted with H^{18}OCH$_3$ in the presence of acid. What would you expect the main product to be and why?**

The product would be ^{18}O labeled Methyl Benzoate. This is because in the acid catalyzed formation of Esters, the Alcohol acts as the nucleophile, while one of the Oxygens of the acid (after protonation) is the leaving group. Therefore the labeled Oxygen from the Alcohol would be incorporated into the final product.

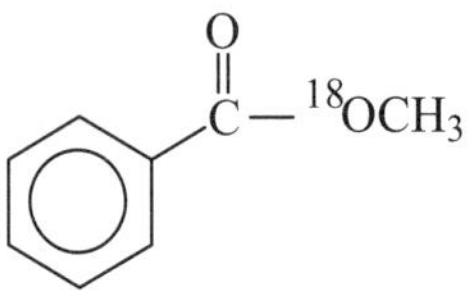

Ph-C(=O)-^{18}OCH$_3$

❏❏ **What is a decarboxylation reaction?**

It is a reaction in which Carbon Dioxide (CO$_2$) is lost from a Carboxylic Acid.

❑❑ **What types of compounds readily undergo decarboxylation?**

Carboxylic acids with a β carbonyl group readily undergo decarboxylation when heated. These compounds include both β ketoacids and β diacids. The temperature at which decarboxylation occurs depends on the structure of the compound.

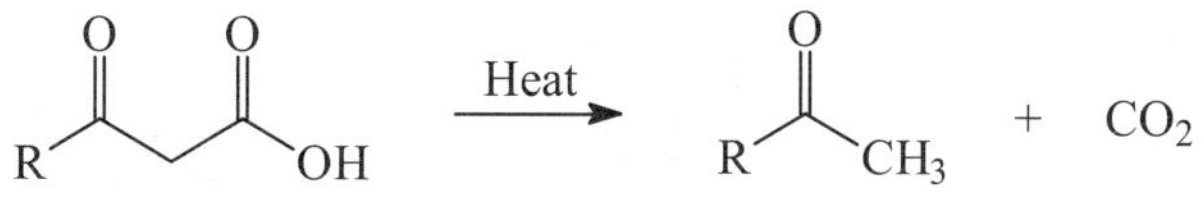

❑❑ **What is the mechanism for the decarboxylation of β ketoacids and β diacids?**

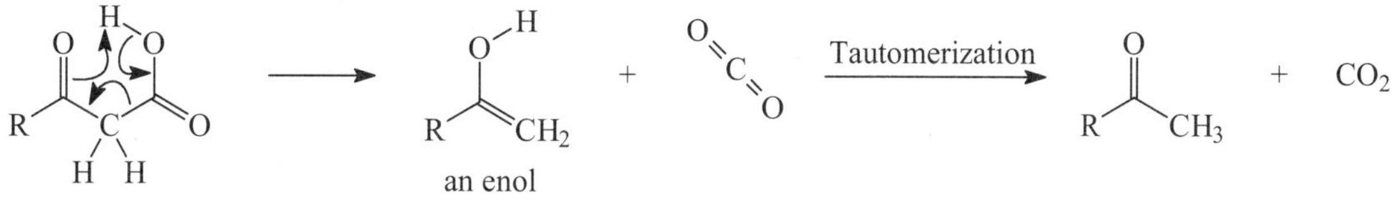

❑❑ **What effect does substitution at Carbon-2 of β ketoacids and β diacids have on the decarboxylation reaction?**

Absolutely none. Decarboxylation proceeds readily with a wide variety of substituted analogs because the Hydrogens on Carbon-2 take no part in the reaction.

❑❑ **What is the mechanism for the acid catalyzed esterification of Carboxylic Acids?**

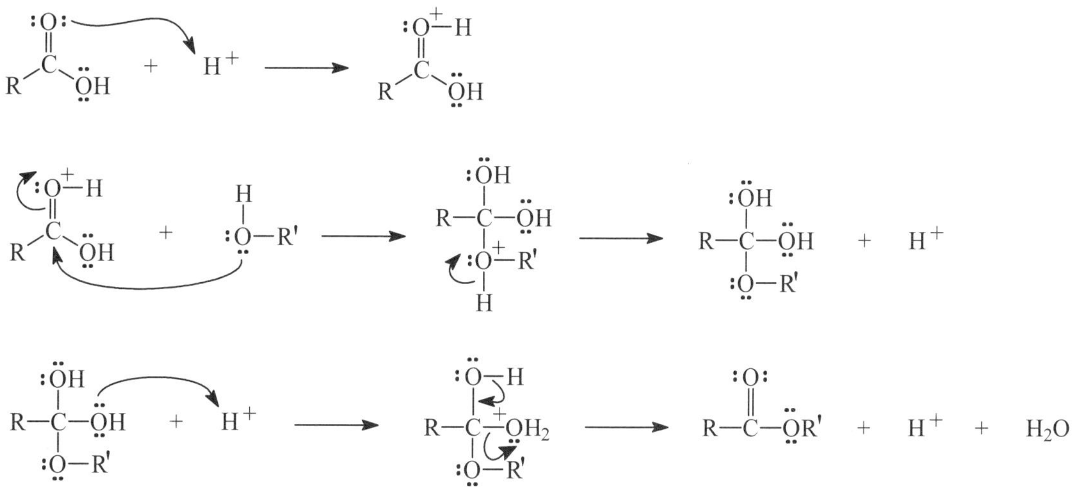

❑❑ **What is the Hell-Volhard-Zelinskii reaction?**

The Hell-Volhard-Zelinskii reaction is a reaction between a Carboxylic Acid and a Halogen (Br_2 or Cl_2) in the presence of a Phosphorus Trihalide. The reaction results in replacement of an α Hydrogen with a Halogen atom. The usual reaction involves Bromine as the Halogen and PCl_3 as the Phosphorus Trihalide.

☐☐ **What is the suspected interemediate in the Hell-Volhard-Zelinskii reaction?**

Carboxylic acids do not readily form enols, which are the usual intermediates in α-halogenation. In the HVZ reaction, the phosphorus trihalide first converts the acid to an Acid Chloride intermediate, which then enolizes and is halogenated in the α position. The halogenated acid halide is converted back to the Carboxylic Acid upon addition of water.

$$R-CH_2-\overset{O}{\underset{\|}{C}}-OH \xrightarrow{PCl_3} \left[R-CH_2-\overset{O}{\underset{\|}{C}}-Cl \rightleftharpoons R-CH=\overset{OH}{\underset{|}{C}}-Cl \right] \xrightarrow[2) H_2O]{1) Br_2} R-\overset{Br}{\underset{|}{CH}}-\overset{O}{\underset{\|}{C}}-OH$$

☐☐ **Why are the products of the Hell-Volhard-Zelinskii reaction very useful?**

The Halogen may be replaced in a nucleophilic addition reaction to synthesize such products as α-amino acids.

☐☐ **What would be the primary organic product expected in each of the following reactions? Give the IUPAC names for each of the products formed.**

a)  PhCH$_2$C(O)OH + CH$_3$CH$_2$OH $\xrightarrow{H^+}$

b) 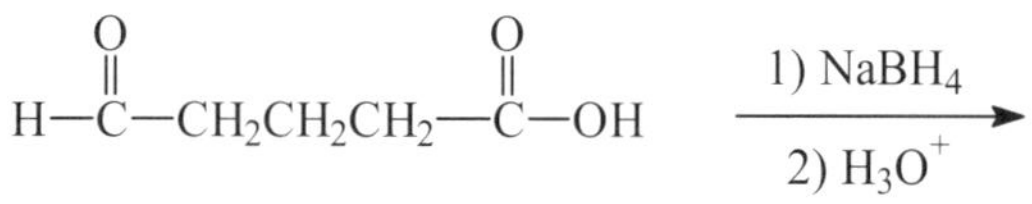 H-C(O)-CH$_2$CH$_2$CH$_2$-C(O)-OH $\xrightarrow[2) H_3O^+]{1) NaBH_4}$

c) H$_3$C-C(O)-O-C(O)-CH$_3$ + PhOH $\longrightarrow$

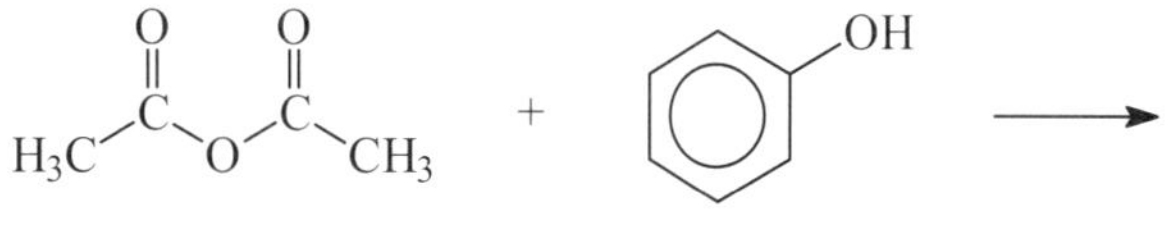

a) 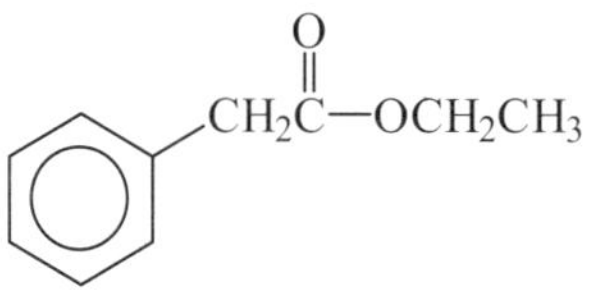 PhCH$_2$C(O)-OCH$_2$CH$_3$

Ethyl Phenylacetate

b) 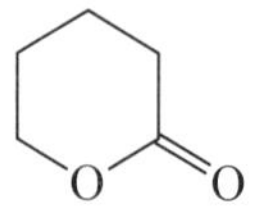 δ-valerolactone structure

5-Pentanolide
(δ-valerolactone)

c) 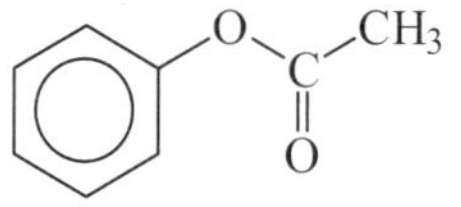 PhO-C(O)-CH$_3$

Phenyl Acetate

❏❏ **What product would be obtained on reaction of Benzoic Acid with each of the following reagents?**

a) NaOH   b) Methanol with catalytic H_2SO_4

c) ThionylCchloride   d) $LiAlH_4$ followed by H_2O

a) 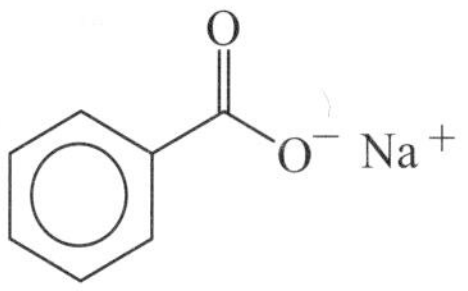 Sodium benzoate (PhCOO⁻ Na⁺)
b) 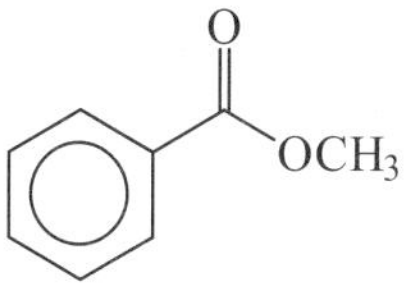 Methyl benzoate (PhCOOCH₃)
c) 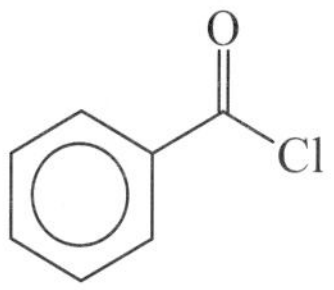 Benzoyl chloride (PhCOCl)
d) 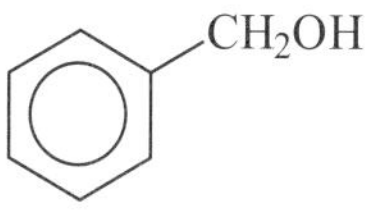 Benzyl alcohol (PhCH₂OH)

❏❏ **What would be the IUPAC standard name of each of the following compounds?**

a) 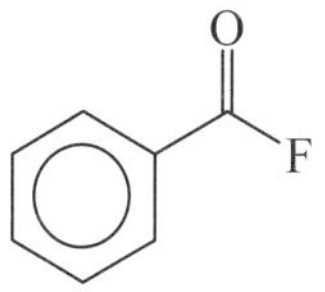 PhCOF

b) 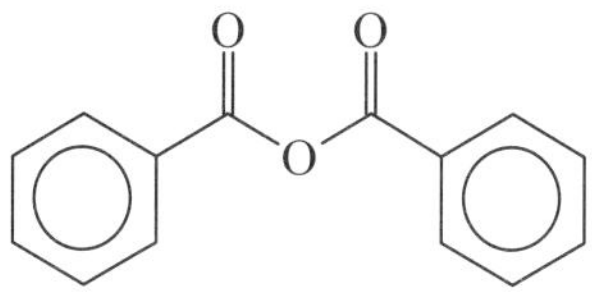 Benzoic anhydride (PhCO-O-COPh)

c) 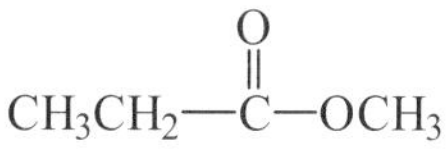 $CH_3CH_2-\overset{O}{\underset{\|}{C}}-OCH_3$

d) 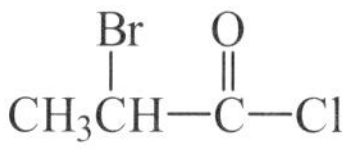 $CH_3\overset{Br}{\underset{|}{C}}H-\overset{O}{\underset{\|}{C}}-Cl$

e) $CH_3C\equiv N$

f) 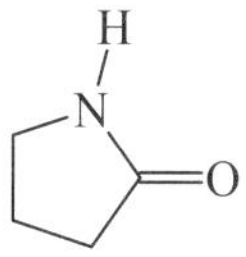 γ-butyrolactam (pyrrolidin-2-one)

(a) Benzoyl Fluoride, (b) Benzoic Anhydride, (c) Methyl Propanoate, (d) 2-Bromo-propanoyl Chloride, (e) Ethanenitrile, (f) 4-Butanolactam (γ-Butyrolactam)

❏❏ **What would be the order of reactivity of each of the following compounds with a nucleophile?**

a) 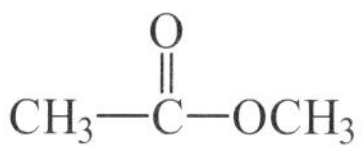 $CH_3-\overset{O}{\underset{\|}{C}}-OCH_3$

b) 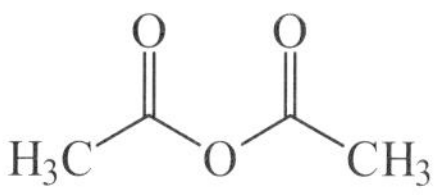 $H_3C-\overset{O}{\underset{\|}{C}}-O-\overset{O}{\underset{\|}{C}}-CH_3$

c) $CH_3-\overset{O}{\underset{\|}{C}}-NH_2$

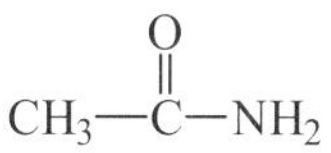

d) 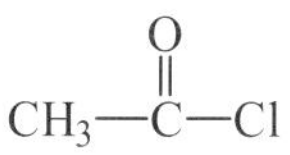 $CH_3-\overset{O}{\underset{\|}{C}}-Cl$

From highest to lowest reactivity: d, b, a, c.

☐☐ **Dimethyl Formamide is a planar molecule with a strong barrier to rotation around the Carbon-Nitrogen bond. Why is this?**

DMF is usually depicted as having the first structure shown below. In this molecular structure, the Nitrogen is tetrahedral (pyramidal) and one would expect free rotation around the C-N single bond. But a second resonance form, in which there is a double bond between the Carbon and Nitrogen, contributes very strongly to the resonance hybrid. This is the contributing structure that results in the observed properties.

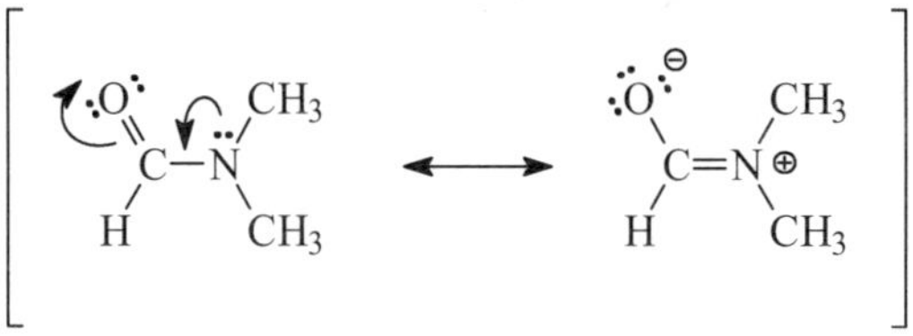

☐☐ **What would be the expected primary organic product in each of the following reactions?**

a) 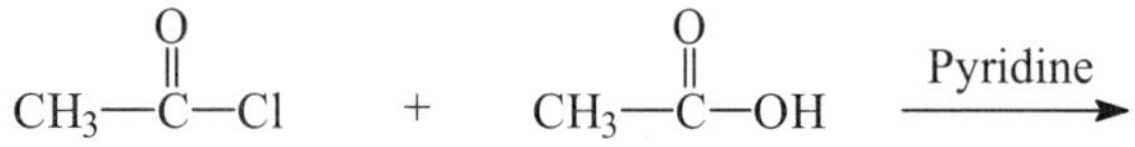 $CH_3-C(=O)-Cl$ + $CH_3-C(=O)-OH$ $\xrightarrow{\text{Pyridine}}$

b) 4-O_2N-C$_6$H$_4$-C(=O)-Cl + piperidine (NH) $\xrightarrow{\text{NaOH} / H_2O}$

c) $Cl-C(=O)-CH_2CH_2-C(=O)-Cl$ + phenol-OH (excess) $\longrightarrow$

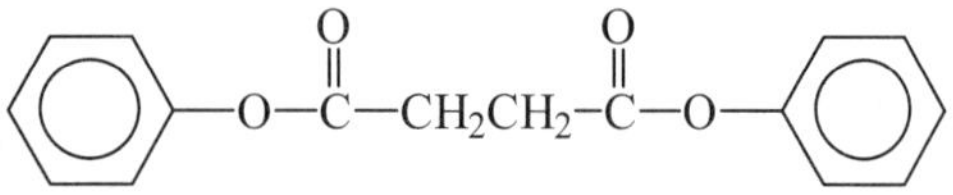

d) 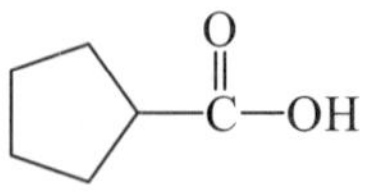 cyclopentyl-C(=O)-Cl + H_2O $\longrightarrow$

a) $CH_3-C(=O)-O-C(=O)-CH_3$

b) 4-O_2N-C$_6$H$_4$-C(=O)-N(piperidine)

c) PhO-C(=O)-CH$_2$CH$_2$-C(=O)-OPh

d) cyclopentyl-C(=O)-OH

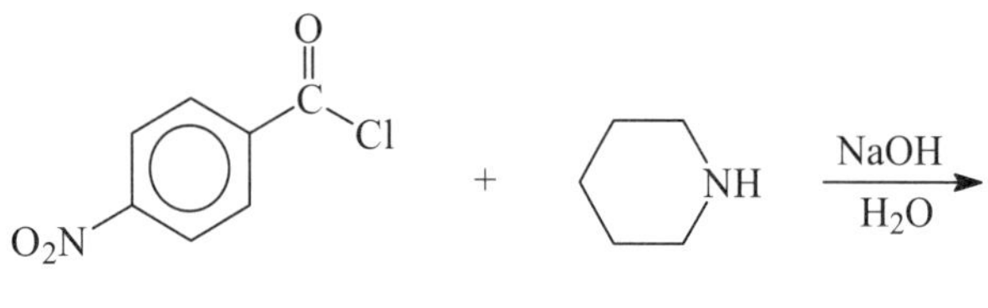

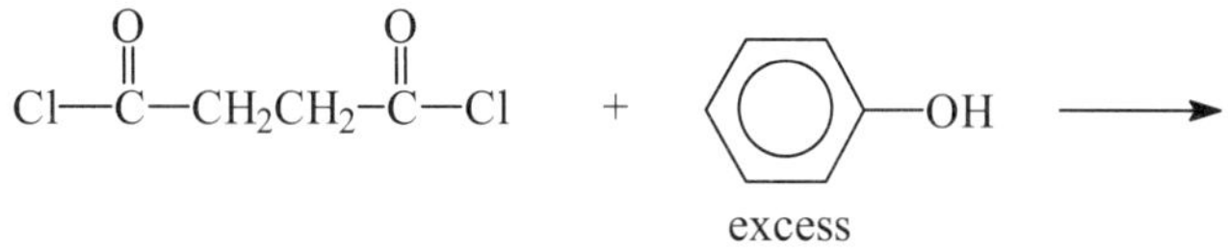

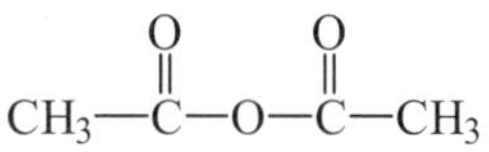

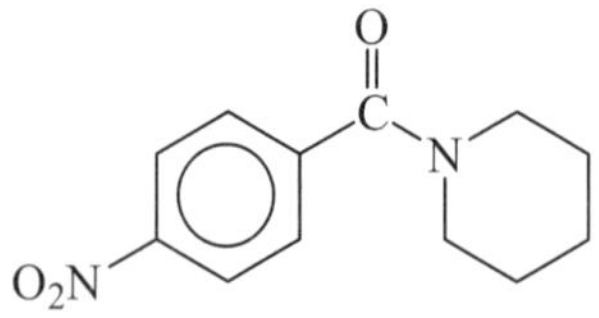

☐☐ **Acid Chlorides readily hydrolyze in water to give Carboxylic Acids. What is the mechanism of this reaction?**

☐☐ **Nucleophilic substitution reactions of Acid Chlorides are much faster (about 1000 times) than substitution reactions of Alkyl Chlorides. Why is this?**

With Alkyl Chlorides, Nucleophilic Substitution reactions involve either a high energy carbocation intermediate (S_N1), or a crowded pentacoordinate transition state (S_N2.) Substitution reactions of Acid Chlorides go through a relatively stable tetrahedral intermediate. This makes for a lower activation energy and an increased rate.

☐☐ **When Acid Chlorides are reacted with Amines to produce Amides, Sodium Hydroxide is usually included in the reaction. Why is this?**

The reaction between an Acid Chloride and an Amine gives HCl as a by-product. If the HCl is not neutralized Sodium Hydroxide, then two equivalents of the Amine are required to complete the reaction because one equivalent must act as a base.

☐☐ **What is the principle reaction of Acid Anhydrides?**

Like Acid Chlorides, acid anhydrides rapidly undergo Nucleophilic Substitution reactions at the carbonyl. A carboxylate ion acts as the leaving group.

☐☐ **What is the Fischer Esterification reaction?**

The Fischer Esterification reaction is the reaction between a Carboxylic Acid and an Alcohol, catalyzed by the presence of acid.

☐☐ **The Fischer Esterification reaction is an equilibrium, and yet Esters are obtained in very good yields. How is this accomplished?**

By removal of water, which is one of the products. This shifts the equilibrium toward products and makes higher yields possible.

☐☐ **Nucleophilic substitution reactions of acid anhydrides, acid halides, Amides, and Esters are all catalyzed by the presence of acid. Why is this?**

In all of these reactions the acid protonates the Oxygen of the carbonyl, making it more susceptible to nucleophilic attack.

❏❏ **Give all of the primary organic products in the following reactions**

a) 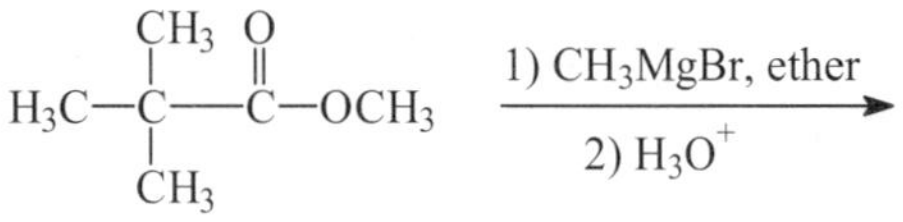 H₃C—C(CH₃)(CH₃)—C(=O)—OCH₃ $\xrightarrow{\text{1) CH}_3\text{MgBr, ether} \quad \text{2) H}_3\text{O}^+}$

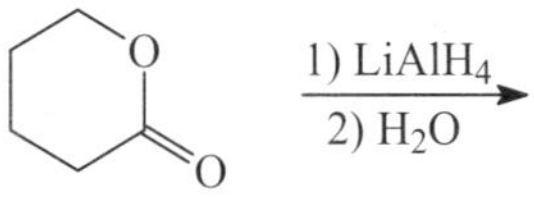

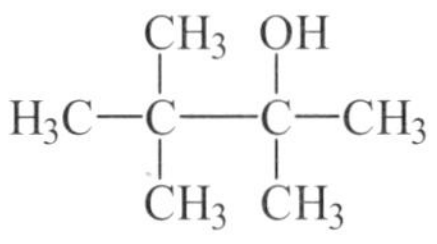

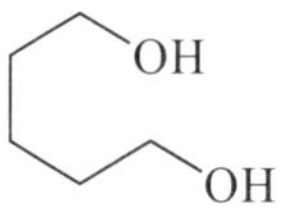

b) δ-valerolactone $\xrightarrow{\text{1) LiAlH}_4 \quad \text{2) H}_2\text{O}}$

c) CH₃O—C(=O)—CH₂—C(=O)—OCH₃ + CH₃I $\xrightarrow{\text{NaOEt}}$

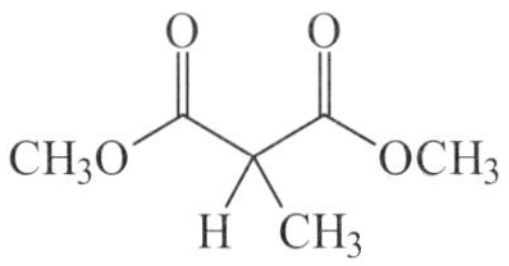

d)  H₃C—C(=O)—OCH₂CH₃ + NH₃ ⟶

a) H₃C—C(CH₃)(CH₃)—C(OH)(CH₃)—CH₃

b) HO—(CH₂)₅—OH

c) CH₃O—C(=O)—C(H)(CH₃)—C(=O)—OCH₃

d) H₃C—C(=O)—NH₂ + CH₃CH₂OH

❏❏ **What is Saponification?**

Saponification is the hydrolysis of Esters under basic conditions.

❏❏ **Why is base catalyzed Ester hydrolysis preferred over acid catalyzed hydrolysis?**

Acid catalyzed hydrolysis is reversible (there is an equilibrium.) Under basic conditions the hydrolysis is irreversible because the Carboxylic Acid that forms as a result of the hydrolysis is converted to a carboxylate salt.

❏❏ **Ammonia, primary Amines, and secondary Amines all react readily with Esters to give Amides. Tertiary Amines, however, do not react. Why is this?**

Tertiary Amines have no proton that can be replaced by an acyl group. So even if a tertiary Amine did attack an Ester, the tetrahedral intermediate would probably simply expel the amino group to regenerate the Ester, because the substitution reaction can't proceed any further.

❒❒ **What are the three most typical ways of preparing an Amide?**

Reaction of a primary or secondary Amine or Ammonia with an Acid Chloride (in the presence of base), an acid anhydride, or an Ester.

❒❒ **What products would you expect in the following reactions?**

a) 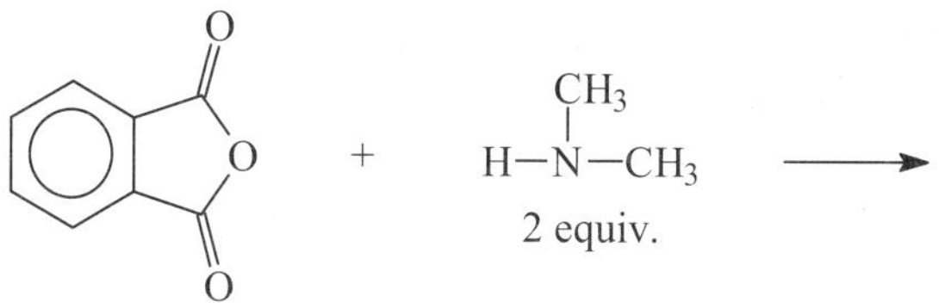
O
O
O
+
CH3
H—N—CH3
2 equiv.

b) 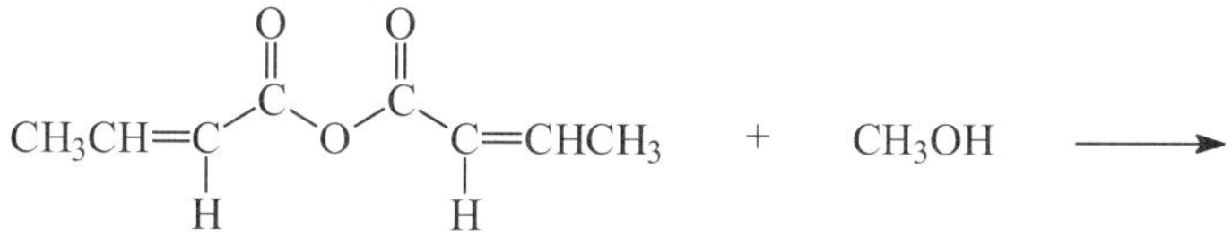
CH3CH=C
C
O
O
C
O
C=CHCH3
H
H
+
CH3OH

c) 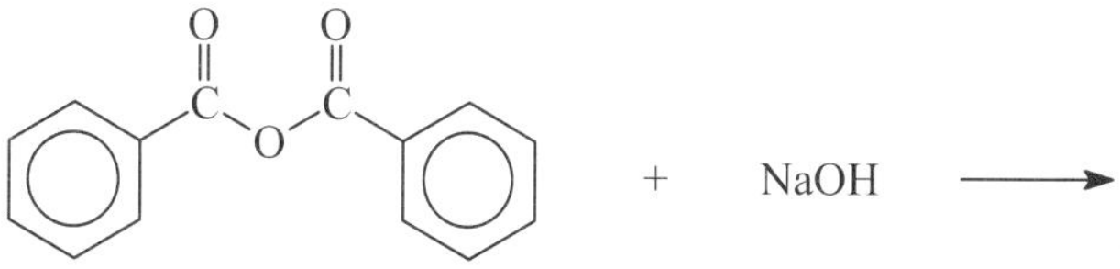
O
C
O
C
O
+
NaOH

d) 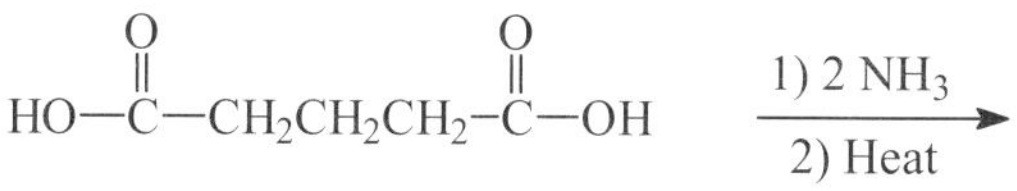
HO—C—CH2CH2CH2—C—OH
1) 2 NH3
2) Heat

a) 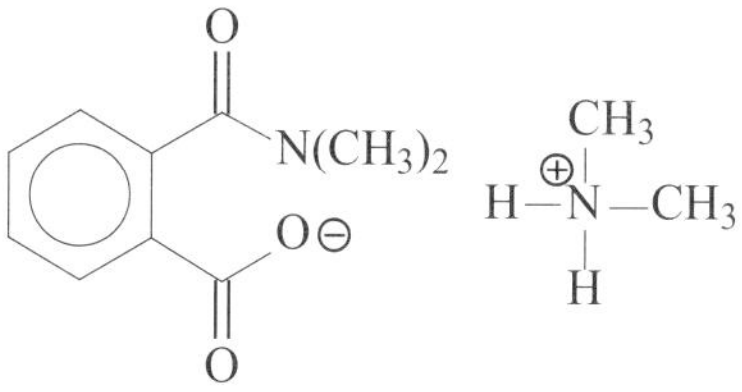
N(CH3)2
O⊖
CH3
H—N—CH3
H

b) 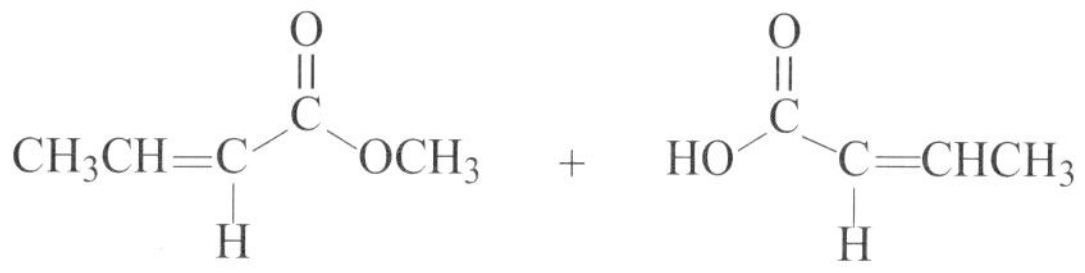
CH3CH=C
OCH3
+
HO
C=CHCH3
H
H

c) 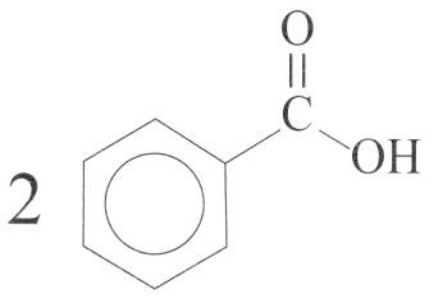
2
OH

d) 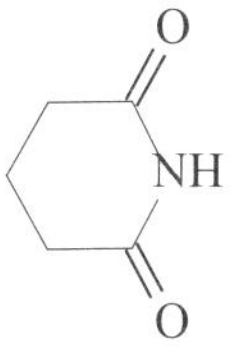
NH

❏❏ **What is the general structure of a Lactam?**

A Lactam is a cyclic Amide. The general structure is:

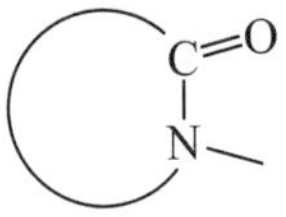

[Structure: ring with C=O attached to N]

❏❏ **Penicillin contains a β-Lactam structure. What is a β-Lactam?**

A β-lactam is a 4-membered ring lactam. The β refers to the non-carbonyl Carbon to which the Nitrogen is attached.

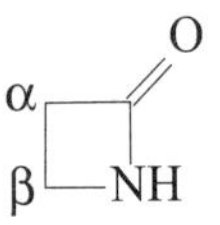

[Structure: 4-membered ring with α carbon, β carbon, C=O, and NH]

❏❏ **Hydrolysis of Amides under both acid catalyzed and base catalyzed conditions is irreversible. Why is this?**

Hydrolysis of an Amide gives a Carboxylic Acid and an Amine (or Ammonia.) Under acidic conditions, the Amine becomes protonated and cannot act as a nucleophile in the reverse reaction, making the process irreversible. Under basic conditions, the Carboxylic Acid is deprotonated, making it unsusceptible to nucleophilic attack. Again, the hydrolysis is irreversible.

❏❏ **Nylon 6, 6 is a polyamide formed by the reaction of Hexanedioic Acid and 1,6-Hexanediamine. What is the repeating structural unit of Nylon 6, 6?**

The structure of Nylon 66 is shown below. The repeating unit is shown in brackets.

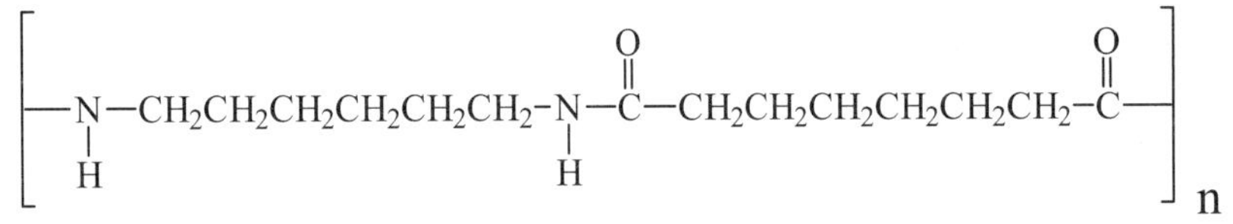

$$\left[-\underset{H}{N}-CH_2CH_2CH_2CH_2CH_2CH_2-\underset{H}{N}-\overset{O}{\underset{\|}{C}}-CH_2CH_2CH_2CH_2CH_2CH_2-\overset{O}{\underset{\|}{C}}- \right]_n$$

❏❏ **The conversion of Nitriles to Ketones is a useful synthetic process. How is this usually accomplished?**

The Nitrile is first reacted with a Grignard reagent to give an imine, which on hydrolysis is converted to a Ketone.

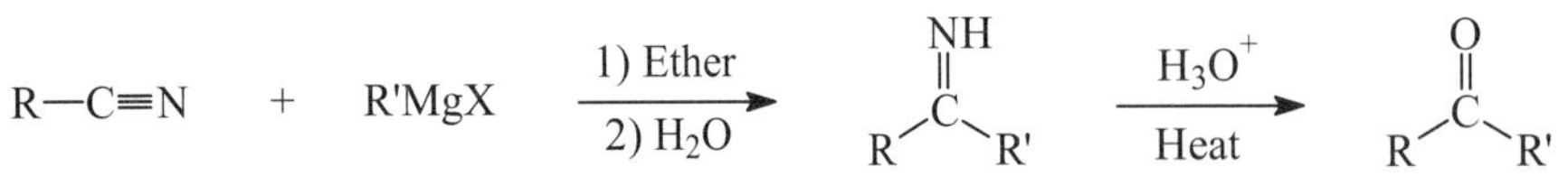

$$R-C\equiv N \; + \; R'MgX \; \xrightarrow[\text{2) H}_2\text{O}]{\text{1) Ether}} \; R-\underset{\|}{\overset{NH}{C}}-R' \; \xrightarrow[\text{Heat}]{\text{H}_3\text{O}^+} \; R-\underset{\|}{\overset{O}{C}}-R'$$

❏❏ **β-Ketoesters can be very useful starting materials in organic synthesis and are typically prepared by a reaction known as the Claisen Condensation. How is this done?**

Esters having alpha protons are treated with an alkoxide base. They undergo self-condensation to give β-Ketoesters on workup.

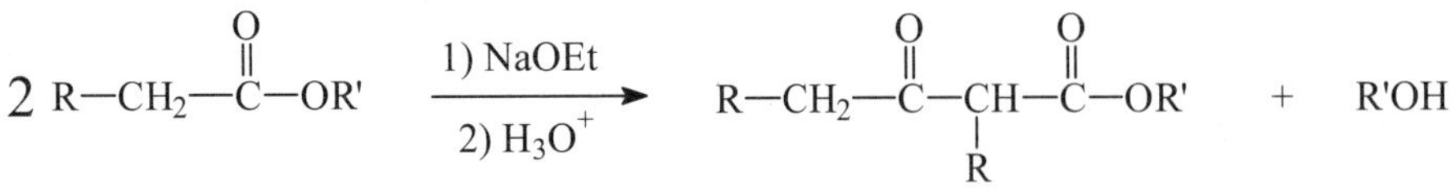

$$2 \; R-CH_2-\overset{O}{\underset{\|}{C}}-OR' \; \xrightarrow[\text{2) H}_3\text{O}^+]{\text{1) NaOEt}} \; R-CH_2-\overset{O}{\underset{\|}{C}}-\underset{R}{CH}-\overset{O}{\underset{\|}{C}}-OR' \; + \; R'OH$$

❏❏ **What is the initial intermediate in the Claisen Condensation?**

The enolate of the starting Ester.

❏❏ **Give the IUPAC names of the following compounds.**

a) 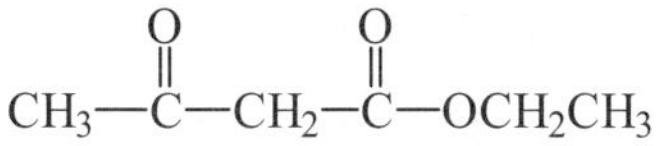 CH₃—C(=O)—CH₂—C(=O)—OCH₂CH₃

b) CH₃CH(CH₃)CH₂—C(=O)—CH₂—C(=O)—O—C₆H₅

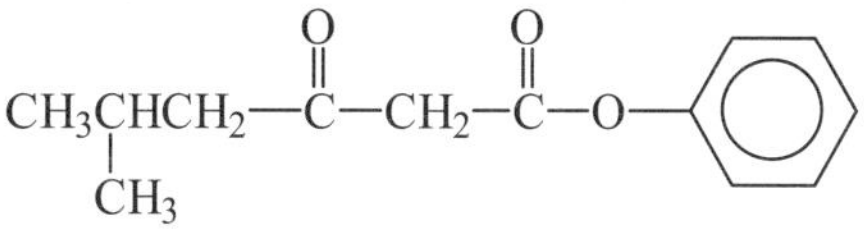

c) CH₃CH₂CH₂—C(=O)—CH(CH₂CH₃)—C(=O)—OCH₃

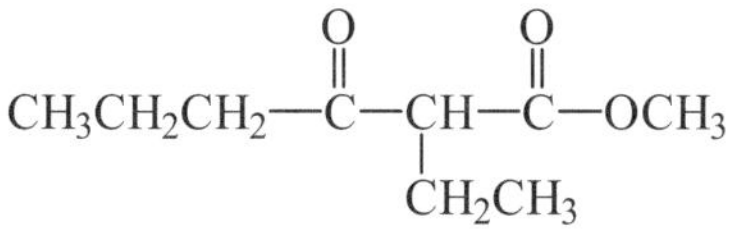

d) 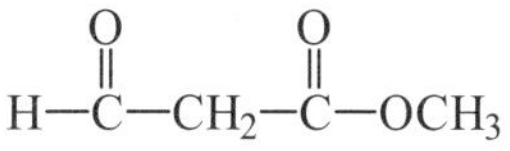 H—C(=O)—CH₂—C(=O)—OCH₃

(a) Ethyl-3-oxobutanoate, (b) Phenyl-5-methyl-3-oxohexanoate, (c) Methyl-2-ethyl-3-oxohexanoate, (d) Methyl-3-oxopropanoate

❏❏ **What is the Acetoacetic Ester synthesis?**

The Acetoacetic Ester synthesis is a reaction used to produce Ketones from Alkyl Halides. Ethyl acetoacetate (Acetoacetic Ester) is deprotonated in the alpha position with an alkoxide base and alkylated with an unhindered Alkyl Halide. The alkylated Ester is then saponified and the resulting β-Keto Carboxylic Acid decarboxylated to give a Ketone.

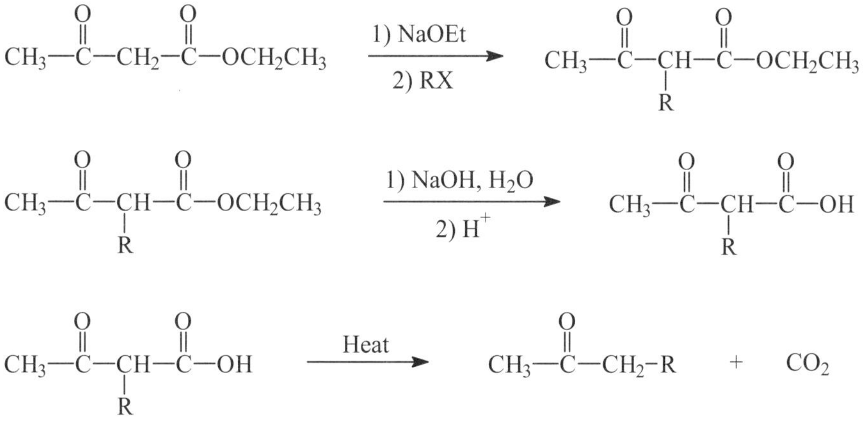

CH₃—C(=O)—CH₂—C(=O)—OCH₂CH₃ →(1) NaOEt, 2) RX)→ CH₃—C(=O)—CH(R)—C(=O)—OCH₂CH₃

CH₃—C(=O)—CH(R)—C(=O)—OCH₂CH₃ →(1) NaOH, H₂O, 2) H⁺)→ CH₃—C(=O)—CH(R)—C(=O)—OH

CH₃—C(=O)—CH(R)—C(=O)—OH →Heat→ CH₃—C(=O)—CH₂—R + CO₂

❏❏ **What structural feature in a Ketone alerts us that it might be prepared by an Acetoacetic Ester synthesis?**

—CH₂—C(=O)—CH₃

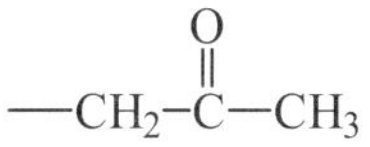

❏❏ **What Alkyl Halide could you use to produce the following molecules by the Acetoacetic Ester synthesis?**

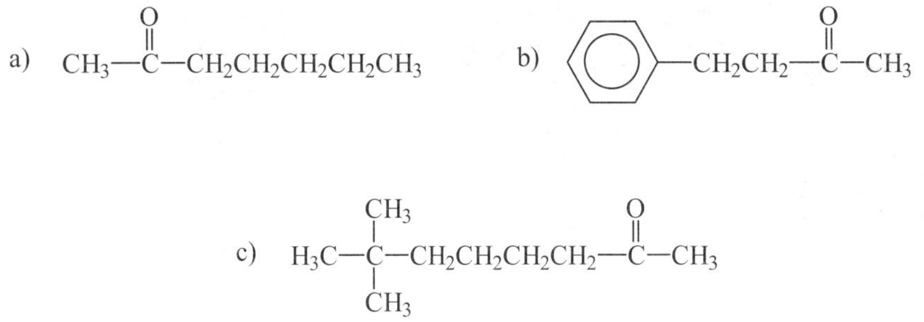

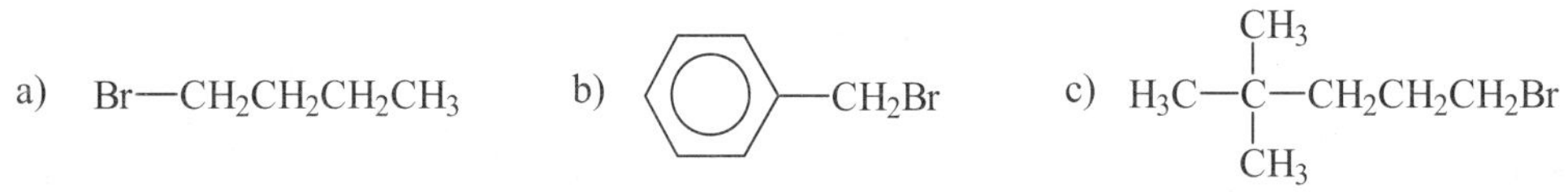

a) CH₃—C(=O)—CH₂CH₂CH₂CH₂CH₃

b) Ph—CH₂CH₂—C(=O)—CH₃

c) H₃C—C(CH₃)(CH₃)—CH₂CH₂CH₂CH₂—C(=O)—CH₃

a) Br—CH₂CH₂CH₂CH₃

b) Ph—CH₂Br

c) H₃C—C(CH₃)(CH₃)—CH₂CH₂CH₂Br

❏❏ **While Sodium Ethoxide suffices to remove the α proton of compounds such as Diethyl Malonate and Acetoacetic Ester, a stronger base must be used to deprotonate Esters that don't produce such stabilized enolates. What is the base of choice in these cases?**

Lithium diisopropylamide (LDA), which has the structure:

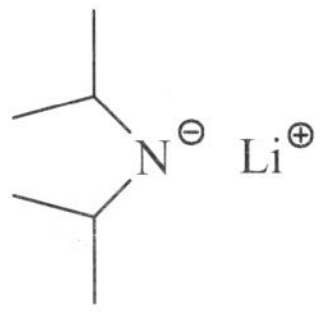

(iPr)₂N⁻ Li⁺

❏❏ **Why is this base used rather than some other strong base?**

It is a strong enough base to remove the protons, but too crowded to act as a nucleophile.

❏❏ **What is the Hofmann rearrangement of Amides?**

If a primary Amide is treated with Bromine or Chlorine in aqueous hydroxide, a rearrangement occurs to give an Amine, in which the carbonyl group of the Amide has been lost, as carbonate, and the Amide Nitrogen is attached to what was the alpha Carbon.

R—C(=O)—NH₂ —[Cl₂, NaOH, H₂O]→ R—NH₂ + Na₂CO₃

❏❏ **What is the stereochemistry of this reaction?**

The reaction occurs with complete retention of configuration at the alpha Carbon.

❏❏ **What is the mechanism of the Hoffman rearrangement of Amides?**

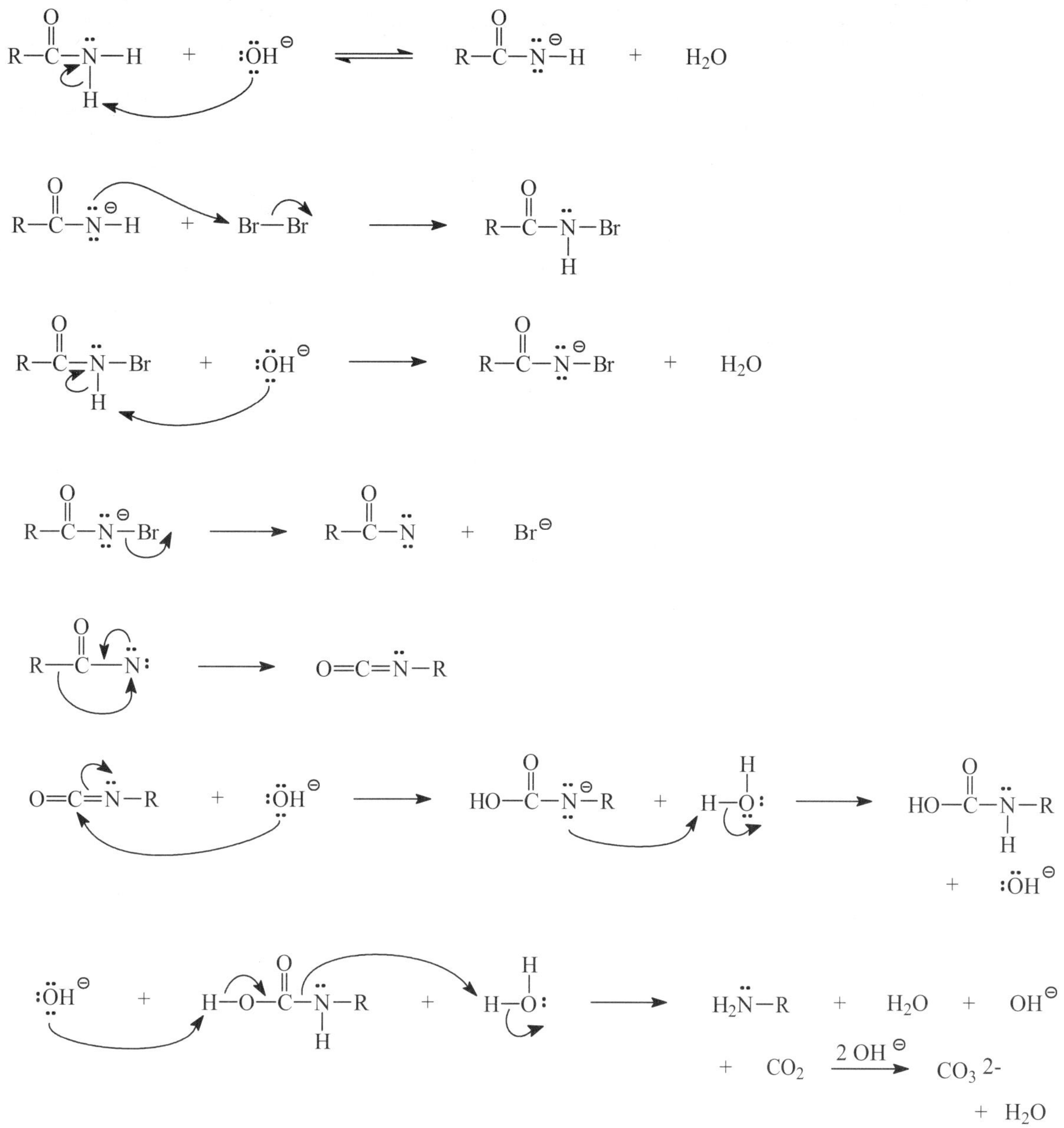

$$R-\overset{O}{\overset{\|}{C}}-\overset{H}{\underset{H}{N}}-H + :\ddot{O}H^{\ominus} \rightleftharpoons R-\overset{O}{\overset{\|}{C}}-\overset{\ominus}{N}-H + H_2O$$

$$R-\overset{O}{\overset{\|}{C}}-\overset{\ominus}{N}-H + Br-Br \longrightarrow R-\overset{O}{\overset{\|}{C}}-\underset{H}{N}-Br$$

$$R-\overset{O}{\overset{\|}{C}}-\underset{H}{N}-Br + :\ddot{O}H^{\ominus} \longrightarrow R-\overset{O}{\overset{\|}{C}}-\overset{\ominus}{N}-Br + H_2O$$

$$R-\overset{O}{\overset{\|}{C}}-\overset{\ominus}{N}-Br \longrightarrow R-\overset{O}{\overset{\|}{C}}-N: + Br^{\ominus}$$

$$R-\overset{O}{\overset{\|}{C}}-\ddot{N}: \longrightarrow O=C=\ddot{N}-R$$

$$O=C=\ddot{N}-R + :\ddot{O}H^{\ominus} \longrightarrow HO-\overset{O}{\overset{\|}{C}}-\overset{\ominus}{N}-R + H-\ddot{O}-H \longrightarrow HO-\overset{O}{\overset{\|}{C}}-\underset{H}{N}-R + :\ddot{O}H^{\ominus}$$

$$:\ddot{O}H^{\ominus} + H-O-\overset{O}{\overset{\|}{C}}-\underset{H}{N}-R + H-\ddot{O}-H \longrightarrow H_2\ddot{N}-R + H_2O + OH^{\ominus} + CO_2 \xrightarrow{2\ OH^{\ominus}} CO_3^{2-} + H_2O$$

❏❏ **What is the balanced equation for this reaction?**

$$R-\overset{O}{\overset{\|}{C}}-NH_2 + Br_2 + 4\ OH^- \longrightarrow R-NH_2 + 2\ Br^- + CO_3^{-2} + 2\ H_2O$$

❏❏ **What is the Malonic Ester synthesis?**

The Malonic Ester synthesis is similar to the Acetoacetic Ester synthesis. A diester of Malonic Acid is converted to its enolate and mono or dialkylated. The substituted product is then hydrolyzed to give the diacid, which decarboxylates on heated. The products are mono or disubstituted acetic acids.

$$CH_3CH_2O-\underset{\underset{H}{|}}{\overset{\overset{O}{\|}}{C}}-\underset{\underset{H}{|}}{C}-\overset{\overset{O}{\|}}{C}-OCH_2CH_3 \xrightarrow[2)\ RX]{1)\ NaOEt} CH_3CH_2O-\overset{\overset{O}{\|}}{C}-\underset{\underset{R}{|}}{\overset{\underset{H}{|}}{C}}-\overset{\overset{O}{\|}}{C}-OCH_2CH_3$$

Diethyl Malonate

$$CH_3CH_2O-\overset{\overset{O}{\|}}{C}-\underset{\underset{R}{|}}{\overset{\underset{H}{|}}{C}}-\overset{\overset{O}{\|}}{C}-OCH_2CH_3 \xrightarrow[Heat]{H_3O^+} R-CH\overset{\overset{O}{\|}}{-C}-OH + CO_2 + 2\ CH_3CH_2OH$$

A Substituted Acetic Acid

❏❏ **How are Nitriles named in IUPAC nomenclature?**

The longest chain containing the cyano group (including the Carbon of the Nitrile) is considered as the parent chain. The name of the Alkane is given the suffix -nitrile. Any substituents are indicated in the usual manner, considering the cyano group Carbon as Carbon one.

$CH_3CH_2CH_2CH_2-CN$

    Pentanenitrile

❏❏ **How are Nitriles named in common nomenclature?**

They are named by taking the common name of the parent Carboxylic Acid, removing the -ic ending, and replacing it with either -nitrile or -onitrile.

    $CH_3CH_2CH_2-CN$

Butyronitrile (from Butyric acid)

❏❏ **Nitriles contain no carbonyl group, and yet they are considered to be Carboxylic Acid derivatives. Why is this?**

Because hydrolysis of Nitriles gives Carboxylic Acids.

❏❏ **How are Nitriles usually prepared?**

There are three common ways of preparing Nitriles: by reaction of cyanides with Alkyl Halides, by reaction of Aldehydes and Ketones with NaCN/acid to give cyanohydrins, and by dehydration of Amides with Phosphorus Pentoxide.

❏❏ **Nitriles are often intermediates in the synthesis of Amines. How is this accomplished?**

Nitriles may be reduced to primary Amines by either catalytic hydrogenation or by reduction with lithium Aluminum Hydride.

❏❏ **What is an imide?**

An imide is a compound in which a Nitrogen is bonded to two acyl groups. Most imides are cyclic.

Succinimide, a cyclic imide

AMINE PEARLS

It's Amino World.
ACS bumper sticker

☐☐ **What is the most significant chemical property of Amines?**

Their basicity.

☐☐ **Why are Amines basic?**

Because the Nitrogen of Amines has an unshared pair of electrons that can be donated to form a covalent bond.

☐☐ **What other chemical property results from the lone pair on the Nitrogen?**

Amines are also nucleophilic.

☐☐ **How are Amines named in IUPAC nomenclature?**

Simple Amines may be named as either Alkylamines or Alkanamines. When named as Alkylamines, the word "amine" is added to the end of the name of the alkyl group attached to the Nitrogen. For purposes of naming the alkyl group, the Nitrogen is considered to be attached to the end of a chain, and any substituents on that chain are indicated by naming them and giving the number of the Carbon to which they are attached. The Carbon bearing the Nitrogen is considered to be Carbon 1. Even alkyl groups attached to Carbon 1 must be named as substituents.

$$\underset{\text{1-Methylbutylamine}}{\underset{\underset{\text{NH}_2}{|}}{\text{CH}_3\text{CHCH}_2\text{CH}_2\text{CH}_3}}$$

1-Methyl substituent — Butyl group (1 2 3 4)

2-Methylcyclopropylamine

To name an Amine as an alkanamine, name the alkyl group as an Alkane and replace the -e ending with -Amine. The position of the Amine is indicated by giving the number of the Carbon to which it is attached. The chain is numbered so as to give the lowest number to the Amine.

$$\underset{\text{2-Propanamine}}{\underset{\underset{\text{NH}_2}{|}}{\text{CH}_3-\text{CH}-\text{CH}_3}} \qquad \underset{\text{4-Methyl-2-pentanamine}}{\underset{\underset{\text{NH}_2}{|}}{\text{CH}_3-\overset{\overset{\text{CH}_3}{|}}{\text{CHCH}_2\text{CHCH}_3}}}$$

Symmetrically substituted dialkyl and trialkyl amines are named by adding the prefix di- or tri- to the name of the alkyl group. Unsymmetrically substituted secondary and tertiary Amines are named as substituted primary Amines. The prefix "N" before the name of an alkyl group indicates that it is substituted on the Nitrogen.

$$\underset{\text{Diethylamine}}{\text{CH}_3\text{CH}_2-\overset{\overset{\text{H}}{|}}{\text{N}}-\text{CH}_2\text{CH}_3} \qquad \underset{\text{N-Ethyl-N-methylpropylamine}}{\text{CH}_3\text{CH}_2-\overset{\overset{\text{CH}_3}{|}}{\text{N}}-\text{CH}_2\text{CH}_2\text{CH}_3}$$

When more than one functional group is present, the -NH$_2$ group may be named as an amino substituent.

❑❑ **What physical property of Amines do people usually notice first?**

Their smell. Volatile Amines smell very much like dead fish.

❑❑ **What is the hybridization of the Nitrogen in Amines?**

The hybridization is sp^3.

❑❑ **Amines, like Alcohols and Alkyl Halides, are classified as 1°, 2° and 3°. How does the classification system of Amines differ from that of Alcohols and Alkyl Halides?**

Alcohols and Alkyl Halides are classified on the basis of the number of alkyl substituents on the Carbon bearing the hetero atom. (O or X), Amines are classified based on the number of alkyl substituents attached directly to the hetero atom (N.)

❑❑ **Would the following Amines be classified as primary, secondary, or tertiary?**

a) HN-CH₃ with CH₃   b) (CH₃CH₂)₃N

c) H₂N-cyclohexyl   d) pyridine

e) N-bicyclic   f) pyrroline NH

(a) 2°, (b) 3°, (c) 1°, (d) 3°, (e) 3°, (f) 2°

❑❑ **How does a Quaternary Nitrogen differ from 1°. 2°, and 3° Nitrogens?**

A Quaternary Nitrogen bears a positive charge and no longer has an unshared pair of electrons. It is therefore no longer basic.

❑❑ **Amines are more soluble in water than similar sized hydrocarbons, but less soluble than comparable Alcohols. Why is this?**

Amines are capable of hydrogen bonding, which makes them slightly soluble in water. But the Nitrogen of Amines is less electronegative than Oxygen, so the hydrogen bonds formed by Amines are not as strong as those formed by Alcohols.

❑❑ **What is the easiest way to estimate the basicity of an Amine?**

By looking at the pKa of the corresponding ammonium salt. A strongly basic Amine will hold on to a proton tightly and the ammonium salt will be less acidic (greater pK_a value.) A weakly basic Amine will not hold the proton very tightly and the ammonium salt will be more acidic (lower pK_a value).

❑❑ **What reaction is used to define the basicity constant of an Amine?**

The basicity constant, K_b, is defined using the reaction in which an Amine acts as a proton acceptor from water.

$R_3N: + H-\ddot{O}-H \rightleftharpoons R_3NH^\oplus + :\ddot{O}-H^\ominus$

❐❐ **What is the formula for the basicity constant of Triethyl Amine?**

$$K_b = \frac{[(CH_3CH_2)_3NH^+][OH^-]}{[(CH_3CH_2)_3N]}$$

❐❐ **What is typical pK_a of an Alkylammonium ion?**

Alkylammonium ions range in pK_a from just below 10 to just above 11.

❐❐ **What is the typical pK_b of an Alkylamine?**

Alkylamines range in pK_b from just below 3 to just above 4.

❐❐ **What is the relationship between the pK_b of an Amine and the pK_a of its Ammonium Salt?**

The relationship is the same as for any conjugate acid/base pair: $pK_a + pK_b = 14$.

❐❐ **How would you arrange the following compounds in order of increasing basicity?**

a) NH_3 b) CH_3NH_2 c) $C_6H_5-NH_2$ d) $(CH_3)_3N$ e) $O_2N-C_6H_4-NH_2$

From weakest to strongest base: e, c, a, b, d.

❐❐ **Why are Aliphatic Amines usually stronger bases than Ammonia?**

The positive charge on the protonated Amine is stabilized by the presence of electron donating alkyl groups because the charge is dispersed over a larger area. This makes the protonated form lower in energy and easier to form, and therefore makes the Amine more basic.

❐❐ **In the gas phase, the order of basicities for Amines is: $NH_3 < 1° < 2° < 3°$. In water solution, however, the order is $NH_3 < 1° \sim 3° < 2°$. Why is there such a difference?**

In the gas phase, an alkyl substituent can stabilize the positive charge of an ammonium ion by donating electrons to the Nitrogen. This increases the basicity of the Amine. More alkyl substituents disperse the charge over a larger area and therefore have a greater effect. In water solution, this type of stabilization still occurs, but the positive charge may also be stabilized by solvation. The positive charge is dispersed when the ammonium ion hydrogen bonds with surrounding water molecules. Primary Amines have two Hydrogens available for H-bonding, while secondary Amines have only one and teriary Amines have none. Thus stabilization from solvation is greatest for primary Amines and Ammonia, and smallest for tertiary Amines. The two stabilization effects in water counterbalance one another. Dialkylamines are more basic in water than either primary or secondary Amines because the ammonium salt has the best combination of stabilization from alkyl groups and stabilization from hydrogen bonding.

❐❐ **Arylamines such as Aniline are quite a bit less basic than alkylamines (about six orders of magnitude.) Why is this?**

In Arylamines the lone pair on the Nitrogen is delocalized into the π system of the aromatic ring. This stabilizes the Amine by decreasing the electron density on the Nitrogen. When the Amine is protonated, this type of stabilization is impossible. In fact, the sp^2 hybridized Carbon of the aromatic ring is electron withdrawing and actually destabilizes the positive charge. These two effects contribute to the decreased basicity, shifting the acid-base equilibrium toward the unprotonated Amine.

❏❏ **How do substituents on the Phenyl ring effect the basicity of Aniline derivatives?**

Electron donating substituents (such as alkyl groups) increase the basicity slightly, usually by less than one pK unit. Electron withdrawing substituents (such as Nitro, Trihalomethyl, or Halo) can decrease the basicity considerably, by two or more pK units.

❏❏ **Would the following para substituents increase or decrease the basicity of Aniline?**

a) —NO_2 b) —C(=O)—CH_3 c) —F

d) —CH_3 e) —C(CH_3)(CH_3)—CH_3 f) —Cl

(a) Decrease, (b) Decrease, (c) Decrease, (d) Increase, (e) Increase, (f) Decrease

❏❏ **3-Nitroaniline has a pK_b of 11.5, while 4-Nitroaniline has a pK_b of 13.0. Why is there such a difference?**

A nitro group has the potential of being electron withdrawing in two different ways. The first is an inductive effect. The Nitrogen of a nitro group has a positive formal charge and this draws electron density through the bonds of the aromatic ring. This type of effect operates no matter where the nitro group is situated on the ring. The other method by which a nitro group can be electron withdrawing is by resonance. This resonance effect is position dependent. If the nitro group is in the para (4) position, the electrons of the amino group can be delocalized onto the Oxygen atoms of the nitro substituent. This is not possible if the nitro group is in the ortho (3) position. The more delocalized the amino electrons, the weaker the base will be, and the higher its pK_b.

Resonance delocalization of the amino electrons is possible when the nitro group is in the 4 position.

No resonance delocalization is possible if the nitro group is in the 3 position.

❏❏ **Guanidine, with a pK_b of 0.4, is the strongest base of all neutral compounds. Why do you think this would be the case?**

H_2N—C(=NH)—NH_2
Guanidine

The protonated form of Guanidine, the Guanidinium ion, is stabilized by resonance. The ion has three resonance forms, which spread the positive charge over all three Nitrogens. This stabilized ion is easier to form, so Guanidine is extremely basic.

❏❏ **Give the IUPAC name for each of the following compounds.**

a) 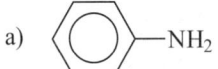   b)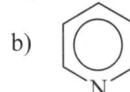

c) 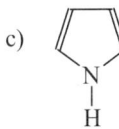   d)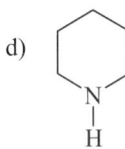

(a) Aniline, (b) Pyridine, (c) Pyrrole, (d) Piperidine

❏❏ **What are heterocyclic and heteroaromatic Amines?**

A hetereocyclic Amine is one in which the Nitrogen is part of a ring system. A heteroaromatic Amine is one where the Nitrogen is part of an aromatic ring.

❏❏ **What would be the IUPAC name for each of the following compounds?**

a) $H_3C-N-CH_3$ with CH_3 below

b) $CH_3CH_2-N-CH_3$ with H below

c) $CH_3CH_2-CH-CH_2$ with NH_2 NH_2 below

d) phenyl-N(CH_3)(CH_3)

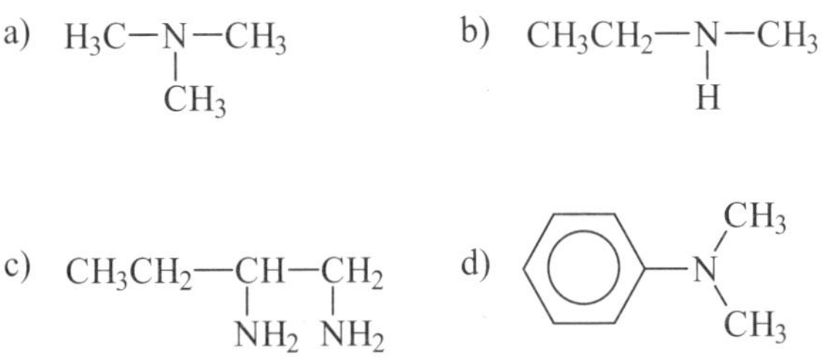

(a) N,N-Dimethylmethanamine or N,N-Dimethylmethylamine, (b) N-Methylethanamine, or N-Methylethylamine, (c) 1,2-Butanediamine, (e) N,N-Dimethylaniline

❏❏ **How would each of the following compounds be named using common nomenclature?**

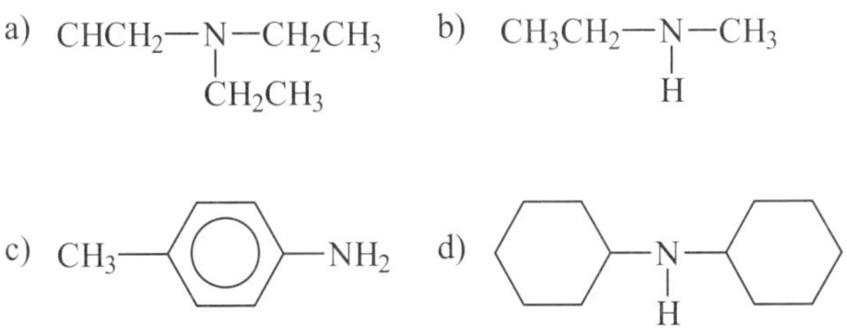

a) $CHCH_2-N-CH_2CH_3$ with CH_2CH_3 below

b) $CH_3CH_2-N-CH_3$ with H below

c) CH_3-phenyl-NH_2

d) cyclohexyl-N(H)-cyclohexyl

(a) Triethylamine, (b) Ethylmethylamine, (c) p-Toluidine, (d) Dicyclohexylamine

❏❏ **What is the geometry around the Nitrogen of an Amine?**

Tetrahedral (the unshared pair occupies one of the vertices). If you only consider the bonded atoms, then it may be described as pyramidal.

◻◻ **An Amine with three different groups attached to the Nitrogen is technically chiral (the lone pair takes the fourth position), but such compounds are usually not optically active. Why is this?**

The pyramidal geometry of the Nitrogen undergoes rapid inversion, with the lone pair flipping from one side to the other. This process quickly racemizes any Chiral Center on the Nitrogen, destroying any potential optical activity.

◻◻ **Would you expect a quaternary ammonium salt with four different alkyl groups to be optically active?**

Yes, because inversion would no longer be possible.

◻◻ **Under what conditions might a non-quaternary Amine with three different groups display optical activity?**

If it was somehow unable to invert as a result of it's structure. Such is sometimes the case in compounds where Nitrogen is the bridgehead atom of a polycyclic system.

This amine would be chiral due to hindered inversion.

◻◻ **What are the accepted names of the following aromatic Amines?**

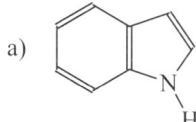
a)

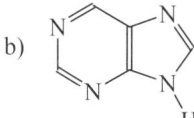
b)

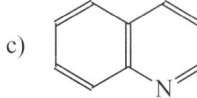
c)

d)

(a) Indole, (b) Purine, (c) Quinoline, (d) Isoquinoline

◻◻ **Alkylation of Ammonia using Alkyl Halides would seem to be a reasonable method for the preparation of alkylamines. Why is this not usually the case?**

Because the Amines resulting from alkylation still have a lone pair available, additional alkylation can occur. Therefore, the result is usually a complex mixture of products. Typically the process is only useful for the production of quaternary ammonium salts, though if a very large excess of Ammonia is used monoalkylated products do predominate.

◻◻ **The basicity of Aniline is low because the lone pair on the Nitrogen is delocalized into the aromatic ring. What effect would you expect if additional phenyl rings were substituted on the Nitrogen?**

The basicity should decrease even more because of additional delocalization. (In fact, the pK_b of Aniline is 9.4, Diphenylamine has a pK_b of 13.2, and Triphenylamine has a pK_b of about 19.)

❑❑ **What are the primary organic products in each of the following reactions?**

a) 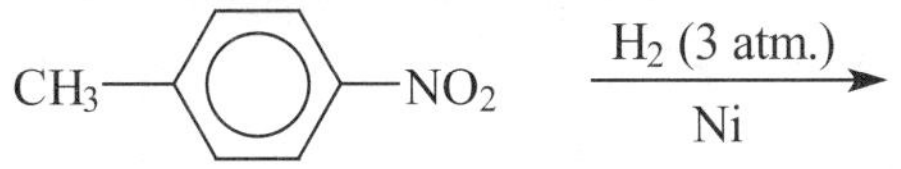 CH₃—C₆H₄—NO₂ $\xrightarrow[\text{Ni}]{\text{H}_2 \text{ (3 atm.)}}$

b)

$\underset{\underset{O}{\diagdown\diagup}}{H_2C—CH_2}$ + NH₃ ⟶

c) Cyclohexyl-CH₂CH₂Br $\xrightarrow[\text{2) NaOH}]{\text{1) NH}_3 \text{ (Excess)}}$

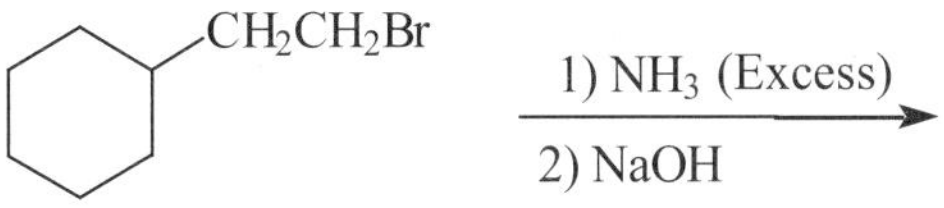

d) CH₃—C≡N $\xrightarrow[\text{2) H}_2\text{O}]{\text{1) LiAlH}_4}$

a)  CH₃—C₆H₄—NH₂   b) 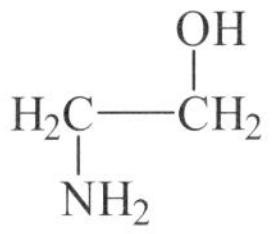 H₂C(NH₂)—CH₂(OH)

c) 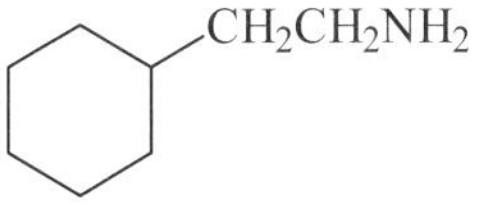 Cyclohexyl-CH₂CH₂NH₂   d) CH₃CH₂—NH₂

❑❑ **Name the following Quaternary Ammonium salts.**

a) 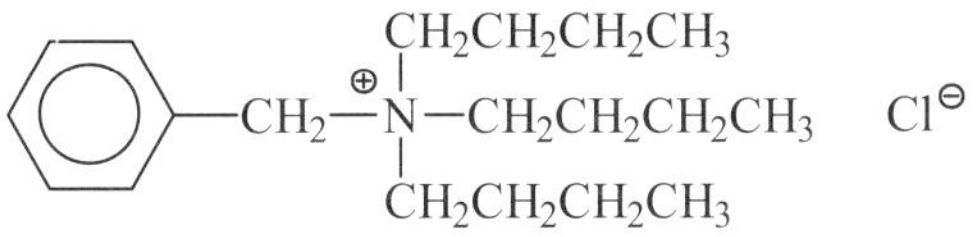 PhCH₂—N⁺(CH₂CH₂CH₂CH₃)₃ Cl⁻

b) 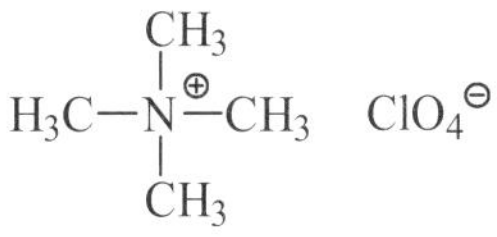 (CH₃)₄N⁺ ClO₄⁻

c) 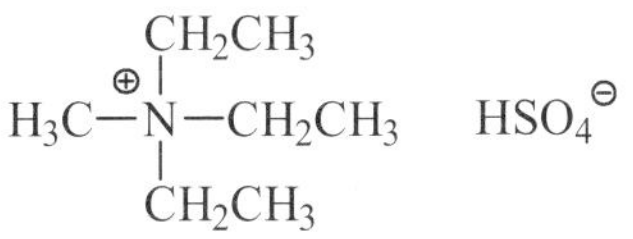 H₃C—N⁺(CH₂CH₃)₃ HSO₄⁻

(a) Benzyltributylammonium Chloride, (b) Tetramethylammonium Perchlorate, (c) Methyltriethylammonium Bisulfate

❏❏ **For what purpose are Tetraalkylammonium salts often used?**

Tetraalkylammonium salts are often used as phase transfer catalysts. Their alkyl groups make the salts soluble in organic solvents, while their ionic character makes them soluble in water.

❏❏ **In the Gabriel synthesis of Amines, Phthalimide is used as a Nitrogen source. What are the basic steps of this process?**

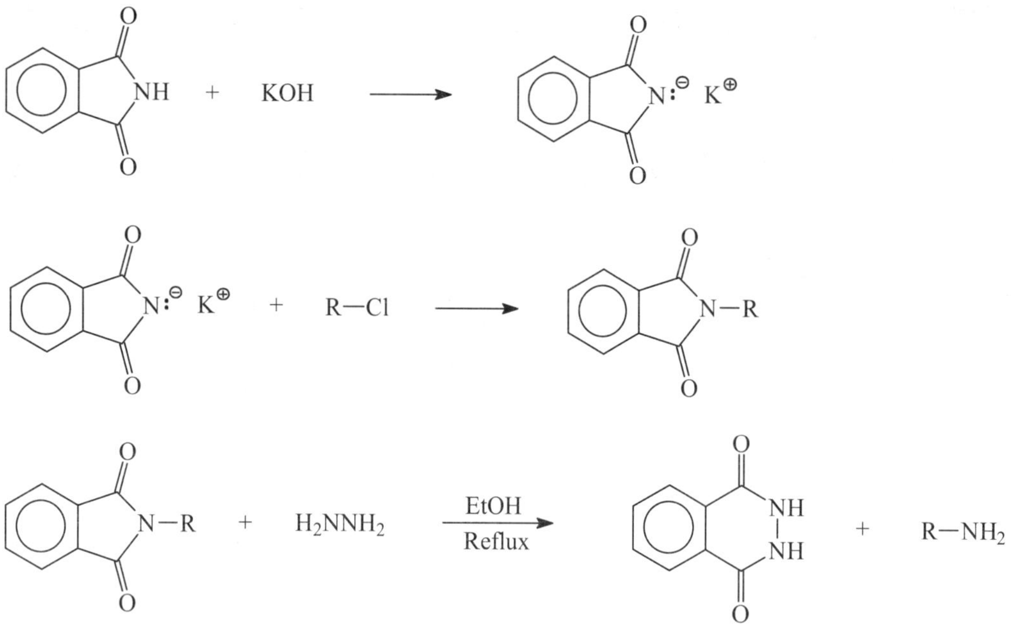

[Reaction scheme: Phthalimide + KOH → Potassium phthalimide]

[Reaction scheme: Potassium phthalimide + R–Cl → N-alkyl phthalimide]

[Reaction scheme: N-alkyl phthalimide + H_2NNH_2 →(EtOH, Reflux) phthalhydrazide + $R-NH_2$]

❏❏ **What are the limitations of the Gabriel synthesis?**

Only primary Amines can be synthesized, and the Alkyl Halide which is used must be 1° or 2°.

❏❏ **What is reductive amination?**

Reaction of an Aldehyde or Ketone with Ammonia or a primary Amine to form an imine, followed by reduction of the imine to give an Amine.

❏❏ **Aldehydes and Ketones can be used to synthesize Amines by first converting them to imines, and then reducing them with an appropriate reagent. $NaBH_3CN$ is often used for this reduction. Why is this?**

Because $NaBH_3CN$ is a strong enough reducing agent to reduce imines, but will not react with carbonyls. So the carbonyl compound, Amine or Ammonia, and the reducing agent can all be mixed in one reaction vessel.

❏❏ **What is the major limitation on reductive amination?**

The amino product can undergo further alkylation by reacting with any remaining carbonyl compound.

❏❏ **How can this limitation be circumvented?**

By using a large excess of the starting Amine or Ammonia.

❏❏ **What reagents are typically used to convert Amides to Amines?**

Typically $LiAlH_4$ is the reducing agent, though H_2 in the presence of a transition metal catalyst has been used.

❏❏ **The Hofmann rearrangement of Amides is a method for converting Amides to Amines. What reagents are used to effect this transformation?**

The Amide is treated with a Halogen (typically Bromine) in the presence of aqueous base. The result is a product in which the alkyl group that was attached to the carbonyl is now bonded to the Nitrogen.

$$R-\underset{\underset{}{\overset{O}{\|}}}{C}-NH_2 \xrightarrow[H_2O]{Br_2, KOH} R-NH_2$$

❏❏ **By what other name is the Hofmann rearrangement of Amides known?**

It is also known as the Hofmann degradation of Amides, because it involves the loss of a Carbon atom from the structure.

❏❏ **Almost any compound containing a Nitrogen may be converted to an Amine by reduction. What classes of compounds are most often used for this purpose?**

Alkyl azides, Nitriles and nitro compounds all give primary Amines upon reduction. Reduction of the carbonyl group of Amides with LiAlH$_4$ provides a route to secondary and tertiary Amines.

$$R-\overset{+}{N}=N=\overset{-}{N}: \xrightarrow{LiAlH_4} R-NH_2 \qquad R-C\equiv N \xrightarrow{LiAlH_4} R-CH_2-NH_2$$
an Alkyl Azide a Nitrile

$$R-NO_2 \xrightarrow[Pt]{H_2} R-NH_2 \qquad R-\underset{\underset{}{\overset{O}{\|}}}{C}-NR_2 \xrightarrow{LiAlH_4} R-CH_2-NR_2$$
a Nitro Compound an Amide

❏❏ **What products are obtained when Amines react with Aldehydes or Ketones?**

Primary Amines react to form imines. Secondary Amines give Enamines upon reaction with Aldehydes and Ketones.

$$H_3C-\underset{\underset{}{\overset{\overset{R}{\diagup}}{\overset{\|}{N}}}}{C}-CH_3 \xleftarrow[1^\circ\ amine]{RNH_2} H_3C-\underset{\underset{}{\overset{O}{\|}}}{C}-CH_3 \xrightarrow[2^\circ\ amine]{R_2NH} H_3C-\underset{\underset{}{\overset{R\diagdown\ \diagup R}{N}}}{C}=CH_2$$
an Imine   an Enamine

❏❏ **What is the Hofmann elimination reaction?**

When a Quaternary Ammonium Hydroxide is heated, it decomposes to give an Alkene, a tertiary Amine, and water.

$$CH_3CH_2-\underset{\underset{CH_2CH_3}{|}}{\overset{\overset{CH_2CH_3}{|}}{\overset{+}{N}}}-CH_2CH_3 \quad OH^- \xrightarrow{Heat} \underset{\underset{CH_2CH_3}{|}}{\overset{\overset{CH_2CH_3}{|}}{N}}-CH_2CH_3 + H_2C=CH_2 + H_2O$$

❏❏ **How is this reaction usually carried out?**

A Quaternary Ammonium Halide is reacted with moist Silver Oxide. Silver Halide precipitates out to leave behind a solution of Quaternary Ammonium Hydroxide. This hydroxide is collected and heated to cause elimination.

❏❏ **What is the Regioselectivity of Hofmann elimination?**

Elimination occurs to preferentially form the least substituted double bond. This reaction does not follow Zaitsev's rule. Reactions of this type are said to follow Hofmann's rule.

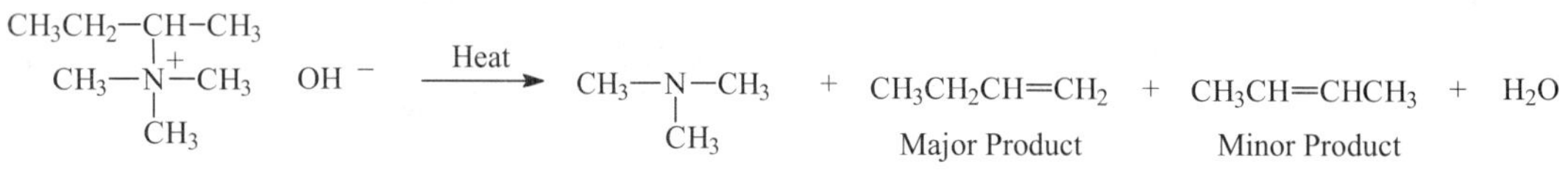

$$CH_3CH_2-CH-CH_3 \atop CH_3-\overset{+}{\underset{CH_3}{N}}-CH_3 \quad OH^- \xrightarrow{\text{Heat}} CH_3-\underset{CH_3}{N}-CH_3 \;+\; \underset{\text{Major Product}}{CH_3CH_2CH=CH_2} \;+\; \underset{\text{Minor Product}}{CH_3CH=CHCH_3} \;+\; H_2O$$

SPECTROMETRY PEARLS

You can observe a lot just by watchin'.
Yogi Berra

❏❏ **What is spectrometry?**

Spectrometry refers to the various instrumental methods for probing the structures and of atoms and molecules

❏❏ **What are the major types of spectrometry used for organic structure determination?**

Nuclear magnetic resonance spectroscopy (NMR), infrared spectroscopy (IR), mass spectrometry (MS), and ultraviolet-visible spectroscopy (UV/Vis).

❏❏ **One of these methods is unlike the others. Which one is it, and how does it differ?**

Mass spectrometry is different from the other methods. All of the other methods involve using electromagnetic radiation to excite molecules, but mass spectrometry does not. Also, mass spectrometry is what's known as a "destructive" method - the sample is destroyed in the process of examination. The other methods are usually non-destructive, and the sample can be recovered intact after examination.

❏❏ **Each of these methods provides a particular type of information. What information does each provide?**

NMR provides information about the backbone of the molecule - the different types of Carbons and Hydrogens and how they are connected. IR gives information about the various functional groups present. MS can provide the atomic mass and atomic formula, as well as some structural information. UV/Vis gives information about the presence, substitution patterns, and conjugation of C-C and C-O double bonds.

❏❏ **How is electromagnetic radiation usually characterized?**

In terms of wavelength and/or frequency.

❏❏ **What does wavelength mean?**

Electromagnetic radiation is transmitted through space in the form of waves. Wavelength, symbolized by λ, is the distance from the crest of one wave to the crest of the next wave.

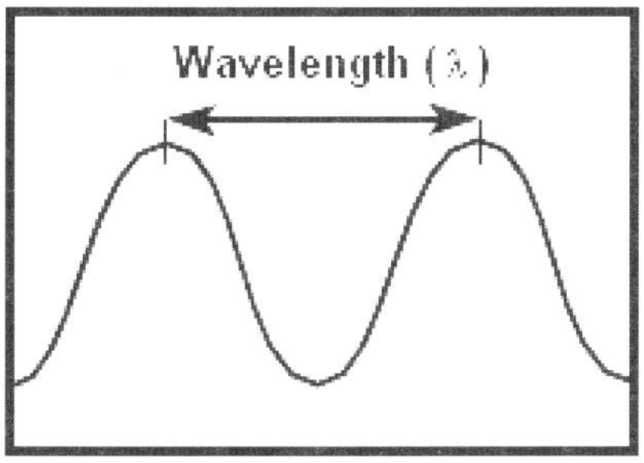

❏❏ What is frequency?

Frequency, symbolized by ν, is the number of cycles per unit time that pass a particular reference point. Frequency is usually measured in cycles per second, also known as Hertz.

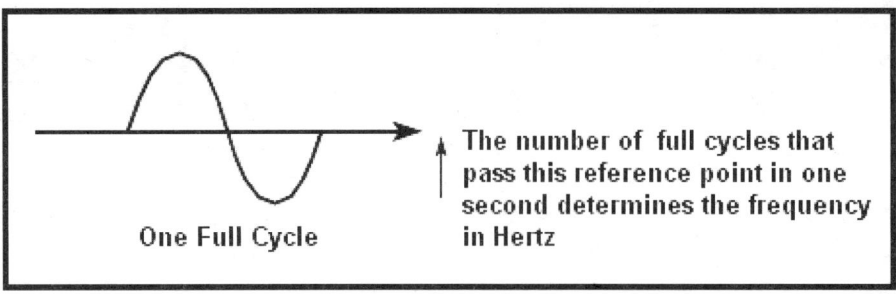

❏❏ What is the relationship between wavelength and frequency?

The numerical relationship is:

$$\nu = \frac{c}{\lambda}$$

Where ν is the frequency in Hertz, c is the speed of light (3×10^{10} cm/sec), and λ is the wavelength measured in cm. Qualitatively, there is an inverse relationship between frequency and wavelength. Since all electromagnetic radiation travels at the same speed, one full cycle of long wavelength radiation takes longer to pass a particular point than does a full cycle of shorter wavelength radiation. Therefore long wavelength radiation has a lower frequency than short wavelength radiation.

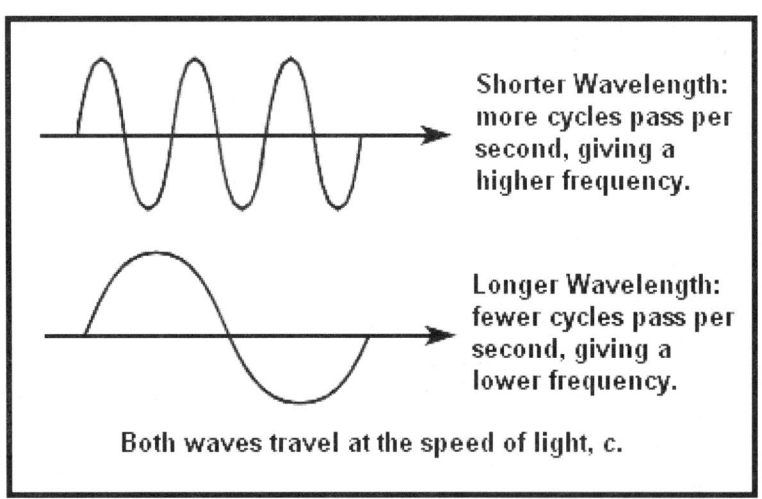

❏❏ What is a photon?

Electromagnetic radiation has both wave properties and particle properties. A photon is a particle of light energy. Photons are also known as quanta.

❏❏ **What is the relationship between the energy of a photon and its frequency and wavelength?**

The relationship is: $E = h\nu = \dfrac{hc}{\lambda}$ where c is the speed of light (3×10^{10} cm/sec), ν is the frequency in Hz, λ is the wavelength in cm, and h is Planck's constant, equal to 9.54×10^{-14} kcal · s/mol.

❏❏ **What are the basic principles of NMR?**

In the absence of a magnetic field, the nuclear spins of atoms are randomly oriented. When certain atoms (such as ^1H or ^{13}C) are placed in a magnetic field, their spins orient themselves either with the magnetic field (parallel to it) or against the magnetic field (antiparallel). Alignment with the magnetic field is a lower energy state. When the nuclei are irradiated with the proper frequency of electromagnetic radiation, the lower energy nuclei (parallel to the field) can absorb energy and the spin can flip to the higher energy state (antiparallel to the field). When this occurs, the nuclei are said to be in resonance with the radiation. As the nuclei relax and return to the lower energy state, they release energy and produce a measurable signal.

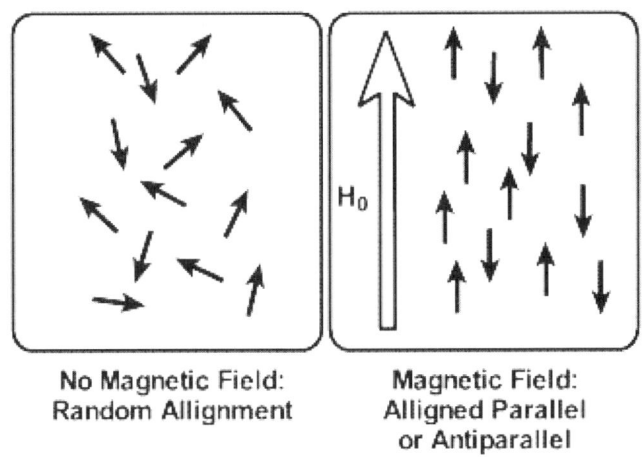

No Magnetic Field: Random Allignment

Magnetic Field: Alligned Parallel or Antiparallel

❏❏ **What is the relationship between the strength of the applied magnetic field and the difference in energy of the two spin states?**

The difference in energy between the parallel and antiparallel spin states increases as the strength of the applied magnetic field (H_0) increases.

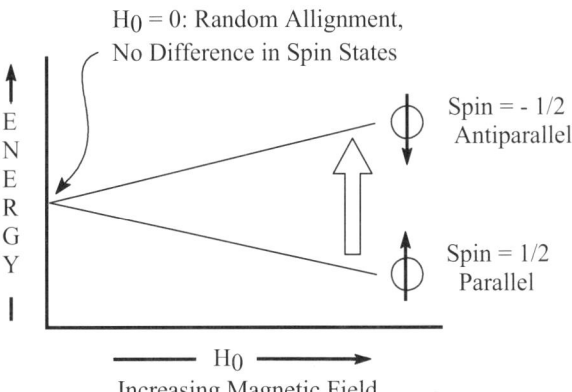

❏❏ **If there are only two possible spin states for the atoms most commonly used for NMR, then why do different nuclei in a molecule go into resonance at different frequencies?**

If all of the nuclei in a molecule were isolated, then each type of atom would absorb at only one particular frequency - all ^{1}H atoms would give one signal, as would all ^{13}C atoms. But these atoms are not isolated. They are surrounded by other atoms and the electrons that bond the molecule together. Electrons in a magnetic field circulate, and by doing so create their own small local magnetic fields. These local fields oppose the applied magnetic field and can shield nearby nuclei. Essentially, different nuclei in a molecule feel magnetic fields of different intensities depending on their environment. The greater the degree of shielding around a nucleus, the lower the magnetic field that the nucleus will feel, and the greater the strength of the applied magnetic field needed to bring the nucleus into resonance. The lower the amount of shielding around a nucleus (or the greater its deshielding), the easier it is to bring into resonance.

❏❏ **What is the structure of a typical NMR spectrometer?**

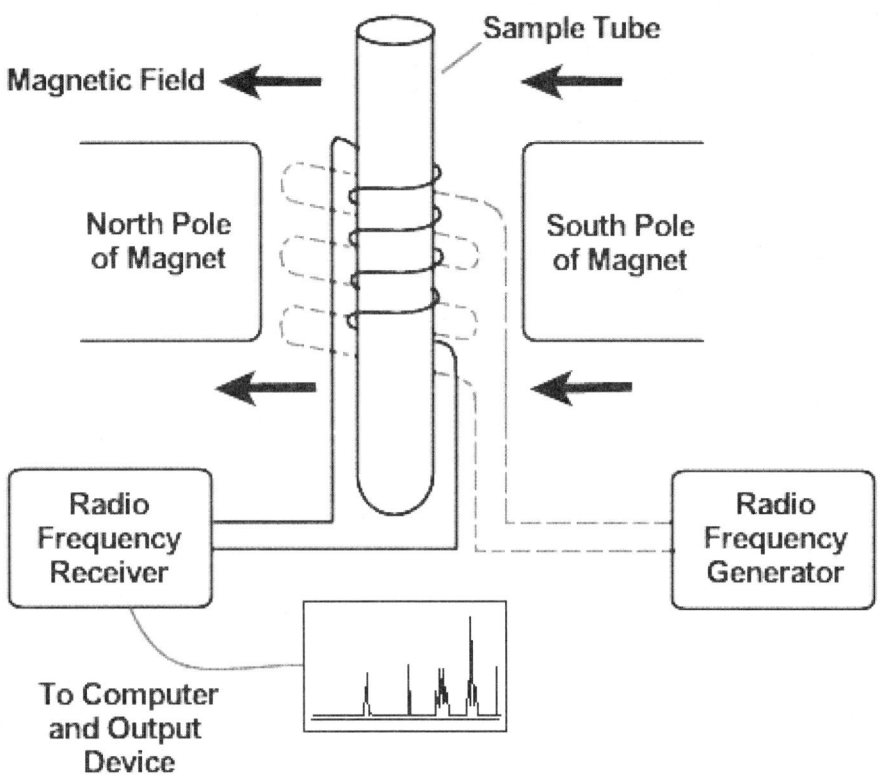

❏❏ **What type of nuclei are "NMR active"?**

Any atom with an odd mass number, an odd atomic number, or both will be potentially useful for NMR.

❏❏ **Why are some nuclei active and others are not?**

All atomic nuclei have a spin quantum number, I. A nucleus with spin quantum number I has exactly $2I+1$ allowed spin states. Nuclei such as ^{1}H and ^{13}C have a nuclear spin quantum number of 1/2, and therefore have $2(1/2) + 1 = 2$ allowed states. These states correspond to the nuclei aligned parallel and antiparallel to an applied magnetic field. Nuclei such as ^{12}C, ^{16}O and ^{32}S have $I = 0$, and therefore have $2(0) + 1 = 1$ allowed spin state, and are therefore not active.

❏❏ **What type of electromagnetic radiation is used to "spin-flip" nuclei?**

Radio-frequency radiation is of the proper energy to effect this transition.

❏❏ **In an NMR spectrum, how are resonance frequencies reported?**

The resonance frequencies in a spectrum are usually reported relative to the resonance frequency of an accepted reference compound. For ^{1}H-NMR and ^{13}C-NMR, the reference compound used is Tetramethylsilane, TMS. In a proton spectrum, the signal from the Hydrogens of TMS is accepted as the zero point of the spectrum. In a Carbon spectrum, the signal from the Carbons of TMS is used as the zero point reference.

❏❏ **What is the structure of TMS?**

$$H_3C-\underset{\underset{CH_3}{|}}{\overset{\overset{CH_3}{|}}{Si}}-CH_3$$

❏❏ **What are δ units?**

δ units are the units typically used to report NMR resonance frequencies. They are expressed as parts per million (ppm) of the applied radiofrequency of the spectrometer. For a spectrometer that uses a radiofrequency of 200 MHz, 1.0 δ (ppm) is equal to 200 Hz.

❏❏ **What is chemical shift?**

Chemical shift, symbolized by δ, is the shift in parts per million (ppm) from the signal of tetramethylsilane (TMS). Chemical shift is an indicator of how shielded an atom is.

❏❏ **In an NMR spectrum, what do the terms upfield and downfield mean?**

Upfield and downfield are relative terms used to compare two signals in an NMR spectrum. Upfield means that the peak is to the right of the one to which it is being compared, while downfield means that it is to the left. A nucleus that has a signal downfield is easier to spin-flip than one that absorbs upfield of it. Downfield nuclei are less shielded (more deshielded) than those that absorb upfield of them.

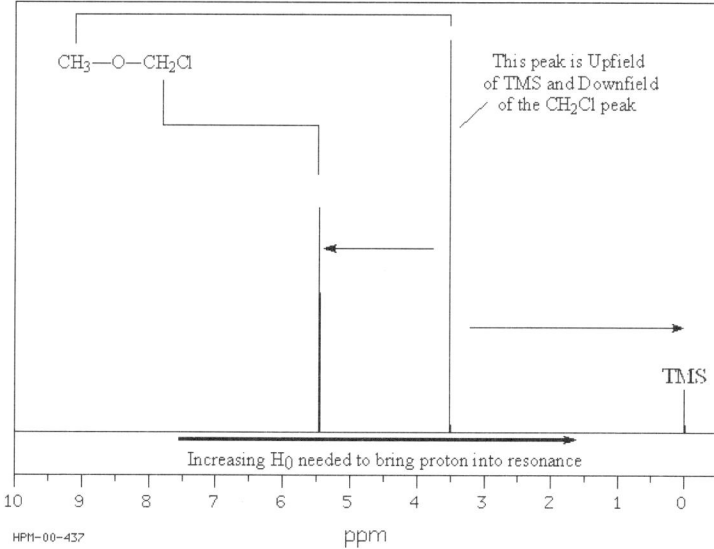

❏❏ **There are two basic types of NMR instrument. What are they and how do they differ?**

The two types of instruments are continuous-wave NMR spectrometers and pulsed Fourier transform NMR spectrometers. In a continuous-wave spectrometer, the sample is irradiated with a constant radiofrequency signal (typically 60 Megahertz) and the strength of the magnetic field, H_0, is varied slightly. As the field changes, the various nuclei in the molecule come into resonance and absorb energy from the radiofrequency signal. The amount of energy absorbed as the field strength varies is measured to give an NMR spectrum. The most common type of NMR instrument these days is a pulsed Fourier transform NMR (FT-NMR.) In these instruments, the magnetic field is held at a constant level and the sample is pulsed with a short, broadband radio frequency signal. This signal flips the spins of all of the nuclei under examination at once. As the nuclei relax to their lower energy state, they each emit a sine wave signal corresponding to their resonance frequency. The intensity of these signals decreases with time. The combined signal from all of the nuclei is resolved into the typical frequency vs. intensity spectrum by a mathematical process known as Fourier transform. A ^1H spectrum can be collected in a few seconds by this technique, so the usual process is to collect several scans and average them to remove noise and increase signal intensity.

❏❏ **What determines the shielding around a nucleus?**

The electron density around a nucleus usually determines how well it is shielded. In an applied magnetic field, electrons generate an induced magnetic field that opposes H_0. This essentially decreases the intensity of the magnetic field felt by nearby nuclei, meaning that a higher applied magnetic field must be used to bring them into resonance. Such nuclei are said to be shielded. Anything that increases the electron density around a nucleus (such as electron donating groups) will increase it's shielding. Anything that decreases the electron density (such as electron withdrawing groups) will decrease the shielding, or deshield the nucleus. Most elements (other than Carbon and Hydrogen) encountered in organic molecules are more electronegative than Carbon and will deshield. Metals or metalloids (such as the silicon in TMS) will be electron donating and shield.

❏❏ **Why are deuterated solvents usually used in proton NMR?**

Most organic solvents contain protons. If these solvents were used for proton NMR, their signals would overwhelm those from any compound dissolved in them. If deuterated solvents are used, the properties of the solvent are retained but there is only a minimal signal from the solvent in the NMR, resulting from a less than complete replacement of the protons by deuterium. Although deuterium is potentially NMR active, its resonance frequencies are nowhere near those of protons.

❏❏ **What is the most common deuterated solvent used in NMR?**

Deuterated Chloroform, $CDCl_3$.

❏❏ **In high field proton FT-NMR spectra, a reference compound is not usually added. How is chemical shift referenced in such cases?**

In proton FT-NMR spectra, chemical shifts are still referenced as though TMS was used to set the zero point, but the position of the scale is actually established by using the small residual amount of protonated impurity left in the deuterated solvent. For example, deuterated chloroform typically contains a small amount of $CHCl_3$, which shows up in the spectrum at δ 7.26. This small peak is found after the FT-NMR spectrum is run, and set to the proper chemical shift, thus adjusting the scale of the rest of the spectrum.

❏❏ **Arrange the following compounds from highest to lowest ^1H chemical shift.**

a) CH_3Cl
b) CH_3I
c) CH_3F
d) CH_3Br

From highest to lowest chemical shift: c, a, d, b.

❏❏ **Why would you expect the chemical shifts of these compounds to be in that order?**

The higher the electron density around a proton, the more shielded it will be, and the further to the right its signal will appear. Halogen atoms are electron withdrawing due to their high electronegativities and deshield nearby protons by removing electron density. The electronegativities (and the electron withdrawing abilities) go in the order F > Cl > Br > I. The protons of Fluoromethane will therefore be the most deshielded and will appear furthest to the left (at the highest chemical shift from TMS), followed by those of Chloromethane, Bromomethane, and Iodomethane.

❏❏ **What is the relationship between chemical shift and Hertz?**

$$\delta = \frac{\text{Shift in Frequency from TMS (in Hertz)}}{\text{Spectrometer Frequency (in Hertz)}}$$

❏❏ **What is the usual chemical shift range for a proton NMR spectrum?**

Typically a proton NMR is scanned from δ 10 ppm to δ 0 (or slightly below). This encompasses the chemical shifts of most types of protons, though Aldehyde protons may appear at about δ 10.1 and Carboxylic Acid protons are known to have chemical shifts as high as δ 13.

❏❏ **In NMR, what is meant by "chemically nonequivalent protons"?**

Equivalent Hydrogens are Hydrogens that have the same chemical environment and give the same ^{1}H-NMR signal. An example of equivalent Hydrogens would be those on a Methyl group. Non-equivalent Hydrogens have different chemical environments and therefore different ^{1}H-NMR signals. Each set of non-equivalent Hydrogens in a molecule will give rise to its own signal in the NMR.

❏❏ **What is the most fool-proof way to test two protons to see if they are chemically equivalent?**

Draw the structure of the molecule in question and replace each one of the protons in turn with a hetero atom (such as Chlorine.) Name each of the structures that you get using IUPAC standard nomenclature. If the names are identical, then the protons are equivalent. If not, then they are non-equivalent.

❏❏ **Is there any other way of recognizing equivalence?**

Yes. Look for symmetry in the molecule. If the molecule has, for example, a mirror plane, then the corresponding protons on either side of the plane will be equivalent. Also, look for protons or groups of protons that can be interchanged by rotation around a C-C single bond.

Mirror Plane: The corresponding protons on either side of the plane are equivalent.

Rotation: All of the protons on to methyl groups are equivalent because they are interchanged by rotation around a bond.

☐☐ **Ideally, how many signals would you expect in the proton NMR spectrum of each of the following compounds?**

a) H—C(CH₃)₃ with structure H—C(CH₃)(CH₃)(CH₃)

b) Toluene (benzene ring with CH₃)

c) CH₃CH₂CH₂CH₂CH₃

d) Cl—CH₂CH₂—C(CH₃)(H)—CH₃

(a) Two, (b) Four, (c) Three, (d) Four

☐☐ **What is spin-spin splitting?**

Spin-spin splitting is the splitting of the NMR signal from a nucleus as a result of the influence of neighboring non-equivalent nuclei.

☐☐ **What is multiplicity?**

Multiplicity is the number of peaks into which the signal from a particular proton is split.

☐☐ **Why does spin-spin splitting occur in ^{1}H-NMR?**

The position of a proton's signal in the NMR depends on the magnetic field felt by the nucleus. Protons on adjacent Carbons can have an effect on this field. If the protons are aligned parallel with the field, their magnetic moments will reinforce it. If they are aligned antiparallels, they will decrease the local magnetic field. Consider the case of a proton on a Carbon adjacent to a methyl group. Each of the protons on the methyl group can have spins that are parallel or antiparallel. There are a total of eight possible combinations of spins. In one combination, all of the spins are aligned with H_0. In three, two of the spins are aligned and one is opposing. In another three one spin is aligned and three are opposing Finally, in one combination all of the spins oppose the applied magnetic field. These eight combinations correspond to four distinct magnetic environments. The signal for the proton next to the methyl group will be split into four peaks corresponding to each of these environments. The intensity of these peaks will ideally be in the ratio of 1:3:3:1, because of the statistical likelihood that each environment will exist in a molecule.

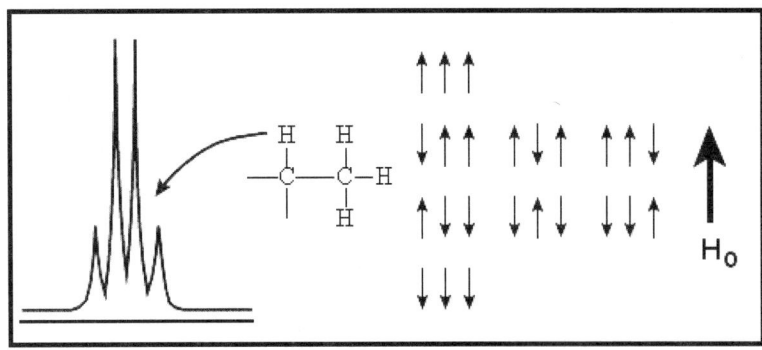

❒❒ **What is an integration line in NMR?**

An integration line is a line that can be added to an NMR spectrum by the instrument that creates it. The vertical rise of the integration line as it travels over a peak is proportional to the area under the peak.

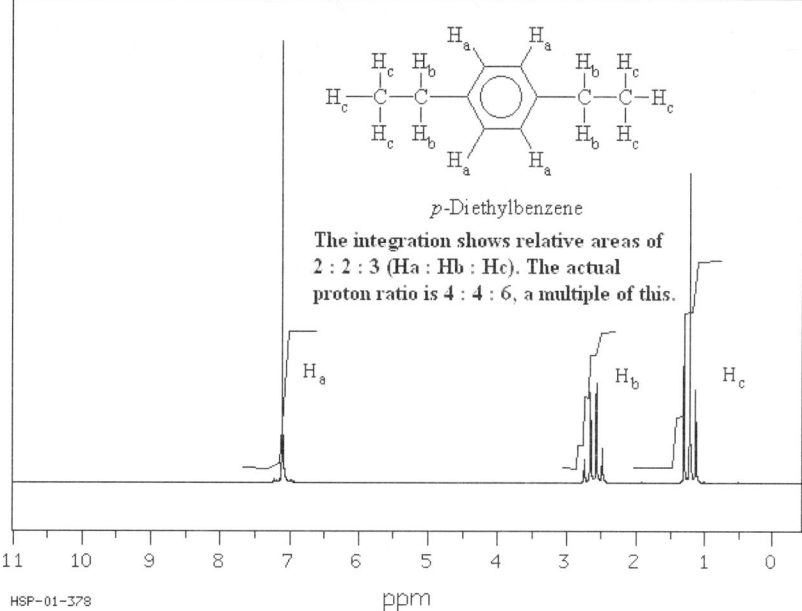

❒❒ **What do the areas of peaks (as measured by an integration line) tell you in a proton NMR spectrum?**

The area under a peak in a proton NMR spectrum is proportional to the number of protons causing the signal. So these peak areas can give you the relative numbers of each type of proton.

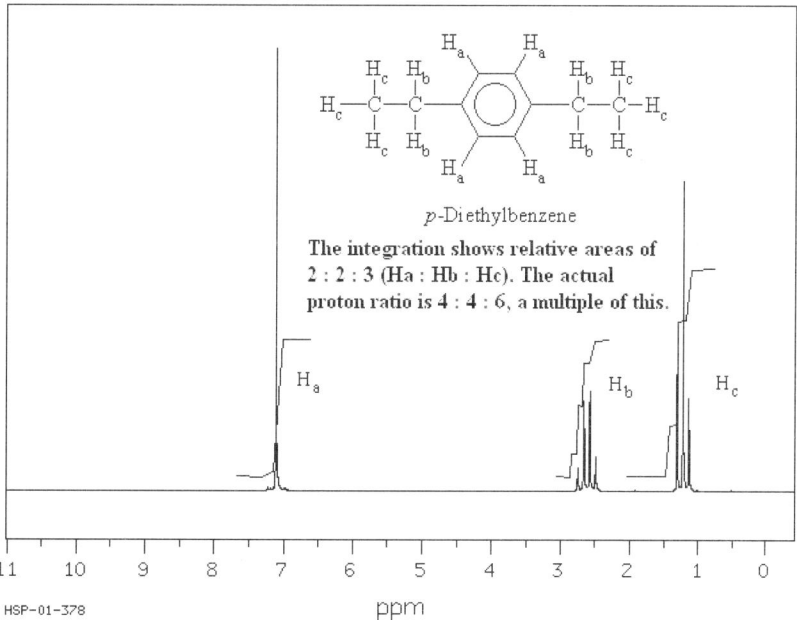

❏❏ **How can you predict the multiplicity of a protons signal based on the structure of a molecule?**

By using the n + 1 rule. The number of peaks into which the signal for a proton will be split due to interaction with a set of n adjacent protons is given by:

Multiplicity = n + 1

In order for a signal to be split, the splitting protons must usually be on an adjacent Carbon (or other H bearing nucleus).

❏❏ **What are diastereotopic protons?**

Diastereotopic protons are protons on the same Carbon that do not share the same chemical environment. As a result, these protons will not have the same chemical shift and may be capable of splitting each other. In many cases, however, the chemical shifts are so similar that this does not occur. Diastereotopic protons may occur in a molecule with a Chiral Center or in a molecule with an unsymmetrically substituted terminal double bond.

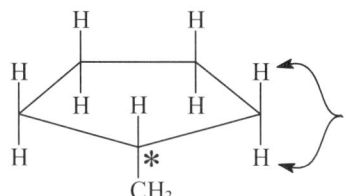

These two protons are in non-equivalent environments because of the chiral center. They are diastereotopic, as are the protons on all the other methylene groups.

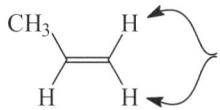

These two protons are diastereotopic as a result of the substitution on the double bond.

❏❏ **What are enantiotopic protons?**

These are protons in mirror-image environments. Enantiotopic protons will have identical chemical shifts and will not split each other.

❏❏ **How are these split signals referred to in NMR?**

They are named based on the number of peaks present.

1 peak: Singlet   5 peaks: Quintet
2 peaks: Doublet   6 peaks: Sextet
3 peaks: Triplet   7 peaks: Septet
4 peaks: Quartet   8 peaks: Octet

❏❏ **What is a coupling constant?**

When a signal is split by an adjacent set of protons, the distance between each of the individual peaks is the same. This distance, measured in Hertz, is known as a coupling constant. The coupling constant between two sets of interacting protons is a measurement of their interaction. The symbol for a coupling constant is *J*.

❏❏ **When the signal for a proton is split, The peaks do not always have the same intensities. How can you most easily predict the intensities of these peaks?**

By constructing Pascal's triangle. In Pascal's triangle, each entry is the sum of the values directly above it to the right and to the left. The entries at each level of Pascal's triangle give the intensity of each of the peaks at that same level of multiplicity. This construct, of course, only gives the ideal intensities.

```
                  1           Singlet
                1   1         Doublet
              1   2   1       Triplet
            1   3   3   1     Quartet
          1   4   6   4   1   Quintet
        1   5  10  10   5   1       Sextet
      1   6  15  20  15   6   1     Septet
    1   7  21  35  35  21   7   1   Octet
```

❏❏ **What happens if a proton interacts with two different sets of adjacent protons?**

The multiplicity may become much more difficult to interpret in the NMR, but can be determined by a tandem application of the n + 1 rule. The first set of n protons splits the signal into n + 1 peaks. The second set of n' protons splits each of these new peaks into n' + 1 peaks. Some of these peaks may overlap. You can determine approximately what such a set of peaks will look like if you draw a splitting diagram (shown below), especially if you know what the coupling constants are.

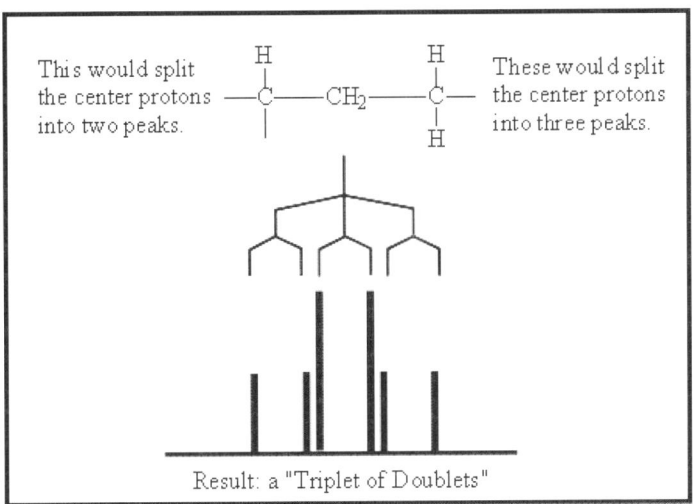

❏❏ **What is the relationship between the size of the coupling constants observed in a spectrum and the spectrometer field strength?**

There is absolutely no relationship. A coupling constant measures only the magnetic interaction between two sets of nuclei. It is independent of NMR field strength.

❏❏ **What can we say about the coupling constants of two sets of interacting protons?**

The coupling constants will be the same. If two sets of protons, A and B are coupled, then both signals will be split with a coupling constant J_{ab}.

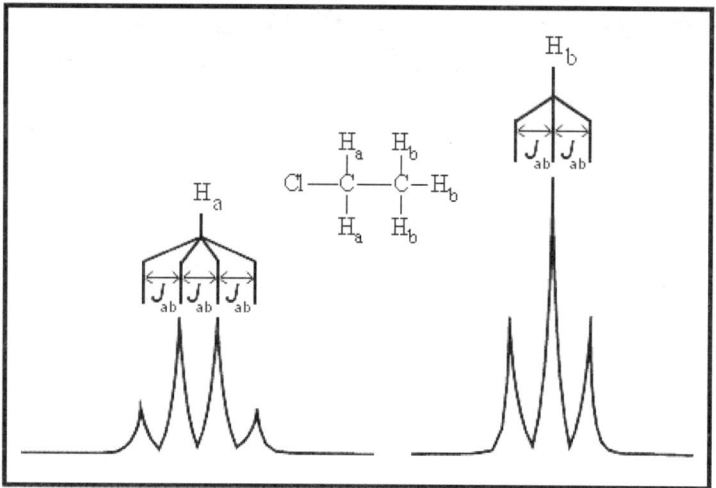

❏❏ **The signals of coupled protons in an NMR are said to "lean" or skew toward each other. What does this mean?**

If two signals are coupled, the peaks of a particular signal on the side closest to the other coupled signal will show slightly enhanced intensity.

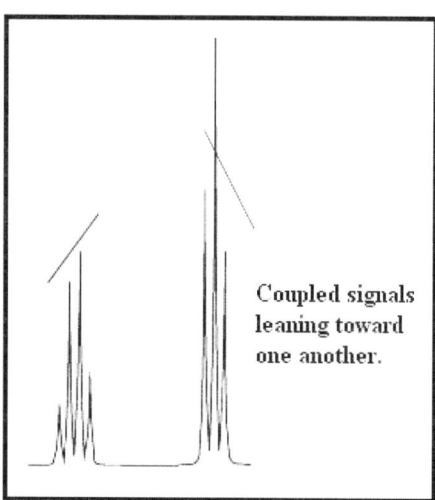

❏❏ **Under what conditions is the skewing of peaks most pronounced?**

The effect increases as the coupled signals get closer in chemical shift.

❏❏ **What is geminal coupling?**

Geminal coupling is coupling (splitting) between two protons on the same Carbon. This can only occur if the protons are diastereotopic. This may be the case if there is a Chiral Center in the molecule, or if the two protons are in non-equivalent positions on an sp^2 hybridized Carbon.

☐☐ **Match the compounds below with their NMR spectra.**

1) 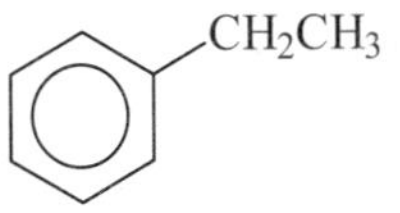 PhCH₂CH₃

2) 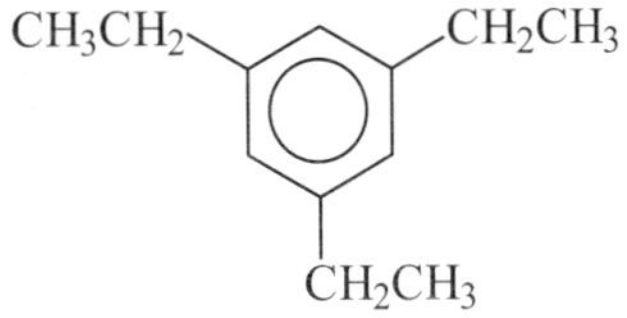 1,3,5-triethylbenzene

3) 1,4-diethylbenzene

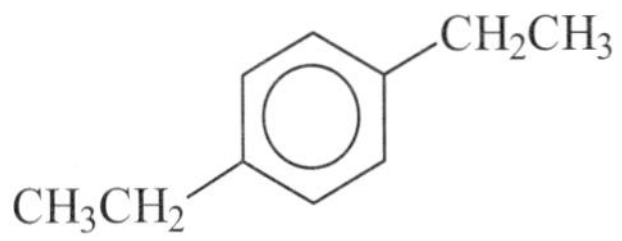

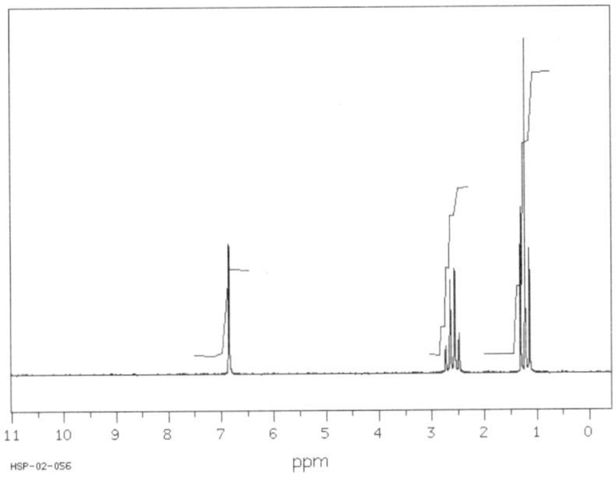

(a)

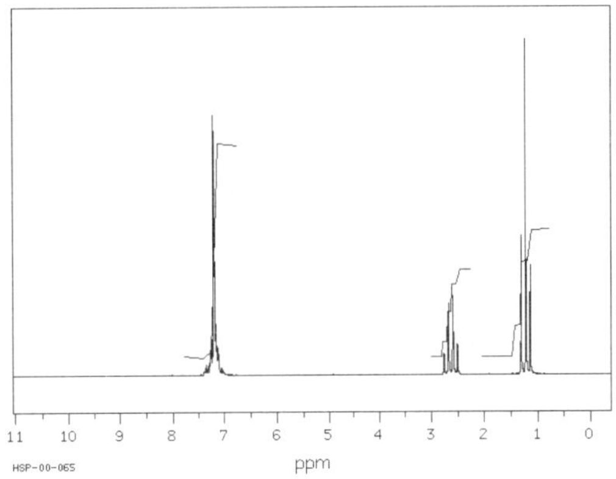

(b)

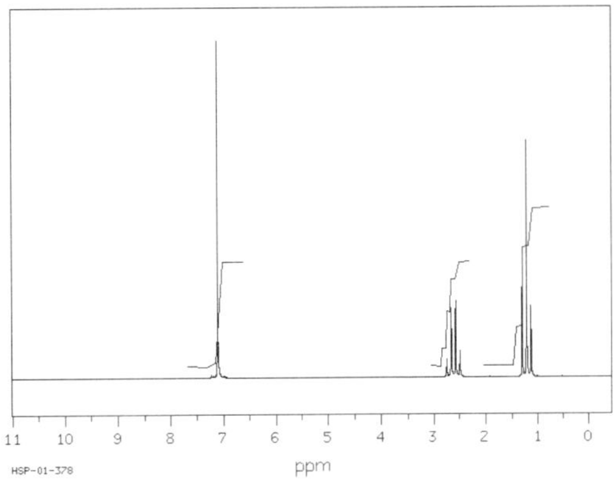

(c)

(1b) (Integration should be 5:2:3), (2a) (Integration should be 1:2:3), (3c) (Integration should be 2:2:3)

☐☐ What splitting patterns would you expect for each of the types of protons on the following molecules?

a) 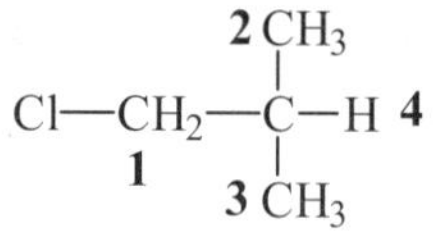 Cl—CH₂—C(CH₃)(CH₃)—H
   1   2 3   4 (2 CH₃ top, 3 CH₃ bottom, 4 H)

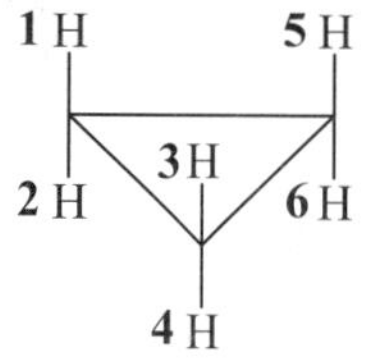

b) cyclopropane with 1H, 2H, 3H, 4H, 5H, 6H

c) HO—C(CH₃)(CH₃)—CH₃
   1   2   3   4

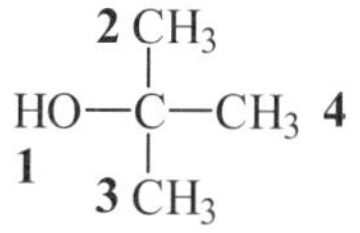

d) 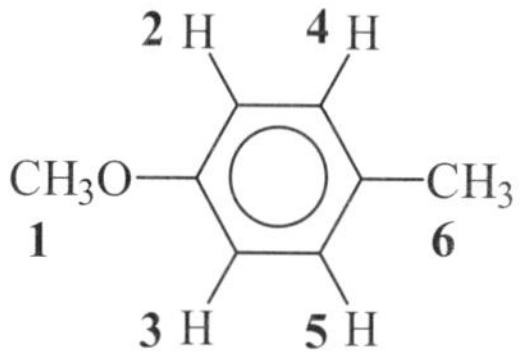 CH₃O—(benzene ring)—CH₃ with 2H, 4H (top), 3H, 5H (bottom), 1 = OCH₃, 6 = CH₃

(a) 1: doublet (split by 4), 2 +3: doublet (they are equivalent sets of H's, split by 4), 4: a "septet of triplets" (split into 7 peaks by 2+3 and split again into threes by 1). (b) 1 -6: singlet (all the protons are equivalent, and so do not split one another.). (c) 1: singlet (O-H protons don't usually couple, but in this case there are no adjacent Hydrogens anyway.) 2-4: all of the methyl protons are equivalent and all will appear as a one singlet. (d) 1: singlet (no adjacent Hydrogens.), 2+3: doublet (these are equivalent and are split by 4 + 5.), 4 + 5: doublet (these are equivalent and are split by 2 + 3.), 6: singlet (no adjacent Hydrogens).

☐☐ Where would you expect the following types of protons to appear in an NMR?

a) 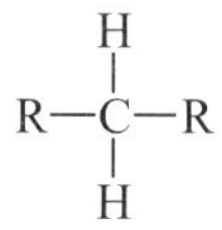 R—CH₂—R

b) 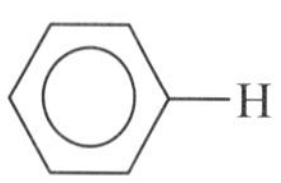 C₆H₅—H

c) 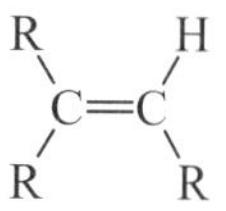 R₂C=CHR

d) R—C≡C—H

e) 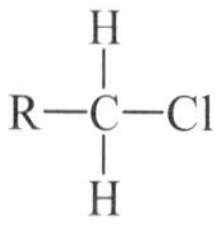 R—CH₂—Cl

f) 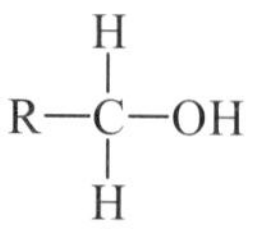 R—CH₂—OH (both types)

g) 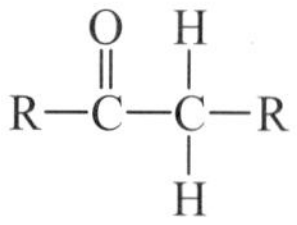 R—C(=O)—CH₂—R

h) 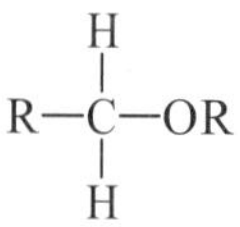 R—CH₂—OR

i) 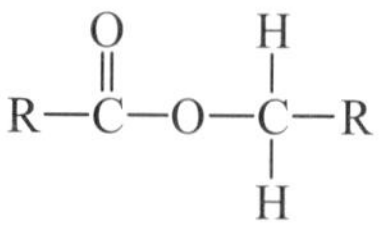 R—C(=O)—O—CH₂—R

(a) δ 1.2 - 1.4, (b) δ 6.5 - 8.5, (c) δ 5.0 - 5.7, (d) δ 2.0 - 3.0, (e) δ 3.6 - 3.8, (f) δ 3.4 - 4.0 (CH₂) and δ 0.5 - 5.0 (OH), (g) δ 2.2 - 2.6, (h) δ 3.3 - 4.0 , i) δ 4.1 - 4.7

❑❑ **Acetylene, Benzene, and Ethylene all have Hydrogens attached to multiply bonded Carbons, yet these protons have incredibly different chemical shifts: the Acetylene protons appear at δ 1.80, the Benzene protons at δ 7.24, and the Ethylene protons at δ 5.25. Why is there such a difference?**

Each of these molecules has π electrons. Charged particles such as electrons, when placed in a magnetic field will tend to circulate. This is the principle of magnetic induction. As these electrons circulate, they create their own local (induced) magnetic field. The exact shape and orientation of the field depends on the geometry of the π system. This field will reinforce the applied magnetic field in some places and oppose it in others. With Acetylene, the proton lies in an area where the induced field opposes the applied field, and so the protons are strongly shielded (and end up at low δ.) With Ethylene and Benzene, the protons lie in an area where the magnetic field is reinforced, deshielding the protons. The effect is stronger for the larger π system of Benzene. This effect, by the way, is also responsible for the high chemical shift of aldehydic protons.

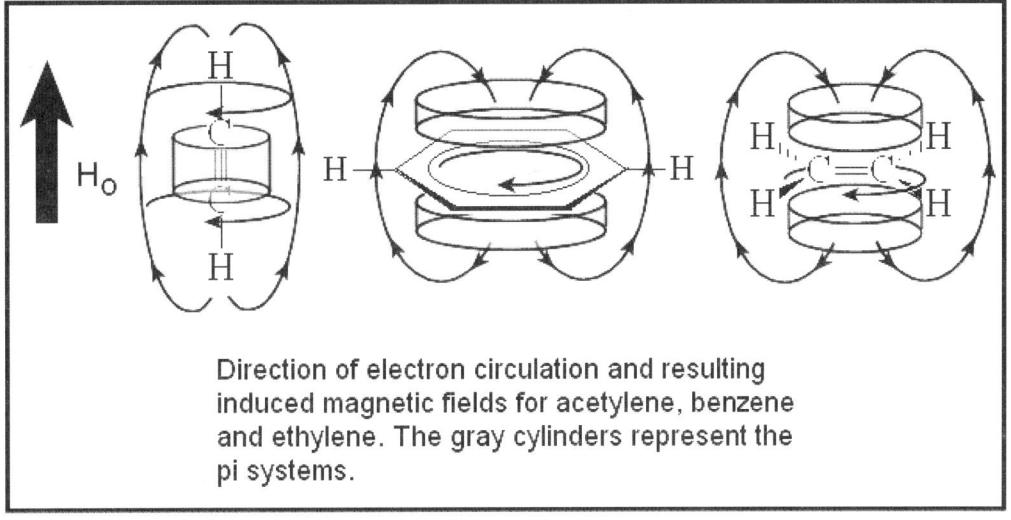

Direction of electron circulation and resulting induced magnetic fields for acetylene, benzene and ethylene. The gray cylinders represent the pi systems.

❑❑ **The development of ^{13}C-NMR took quite a bit longer than that of proton NMR. Why was this?**

Because of the low abundance of ^{13}C - only 1.1 %. This gives a very weak signal. The magnetic moment of ^{13}C is also much smaller than that of ^{1}H, which reduces the signal even more.

❑❑ **There are usually two different ways of running a ^{13}C-NMR, in hydrogen-coupled or hydrogen-decoupled mode. What is the difference between these two?**

The natural abundance of ^{13}C is so low that it is very unlikely that two ^{13}C molecules will be adjacent in a molecule, so there is little chance of these nuclei splitting one another's signals. However, ^{1}H is very common, and therefore under normal circumstances the signals of Carbon nuclei would be split by the Hydrogens bonded to them. This way of running a spectrum is known as hydrogen-coupled mode. It makes it relatively easy to spot groups such as methylenes, methyls, etc. In hydrogen-decoupled mode, the sample is irradiated with two different radio frequencies - one is a broad band that causes all of the Hydrogen nuclei to flip, the other is used to probe the ^{13}C transitions. The broad band radio frequency causes the ^{13}C nuclei to see essentially a constant spin for all of the Hydrogens. Therefore there is no splitting and all of the Carbons appear as singlets.

❑❑ **What is the usual range for a ^{13}C-NMR spectrum?**

Usually about 0 - 220 ppm.

❑❑ **What is the major benefit of such a wide scale?**

It is very unlikely that two non-equivalent Carbons in a molecule will have the same chemical shift. Therefore it is easy to see exactly how many types of Carbons there are.

❒❒ **Why can't integration be used in ^{13}C-NMR to determine the numbers of each type of Carbon?**

In ^{13}C-NMR, the integration is not proportional to the number of Carbons giving the signal. This is because Carbon nuclei, after being flipped, do not all relax back to the low energy state at the same rate. In FT-NMR, a spectrum is collected through a series of scans with a small wait time between them. Those ^{13}C nuclei that relax quickly will have just as many nuclei ready to be flipped in the second scan as in the first, and will give the same signal. Those that relax slowly will still have many nuclei still flipped from the last scan, and so will not absorb as strongly. Peak intensities in ^{13}C-NMR therefore show more about the ability of certain nuclei to relax than it does about the number of nuclei in the molecule producing the signal.

❒❒ **What type of nuclei relax the slowest in ^{13}C-NMR?**

Those that have no protons. These may be recognized by the very low intensity of their peaks.

❒❒ **Where would you expect the following types of Carbons to appear in ^{13}C-NMR?**

a)  Ph—C

b) R—C≡C—R

c) R—C(=O)—OR

d) R—CH$_2$—R

e) R—CH$_3$

f) R—CH$_2$Cl

(a) 110 - 175 ppm, (b) 65 - 90 ppm, (c) 160 - 180 ppm, (d) 15 - 40 ppm, (e) 0 - 40 ppm, (f) 35 - 80 ppm

❒❒ **Assign the peaks in the following ^{13}C NMR spectrum.**

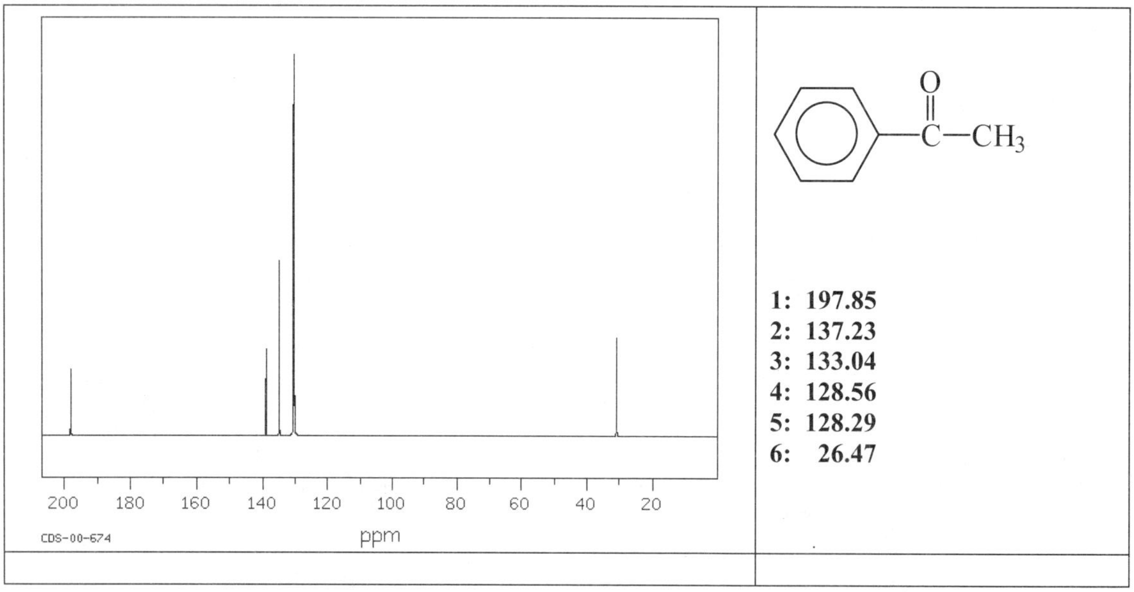

1: 197.85
2: 137.23
3: 133.04
4: 128.56
5: 128.29
6: 26.47

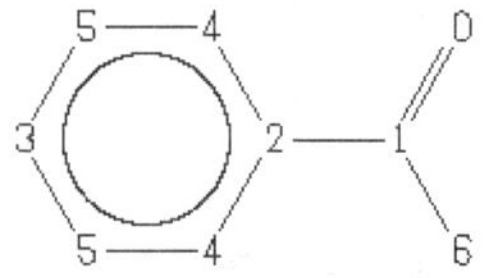

☐☐ **Assign the peaks in the following ¹³C NMR spectrum.**

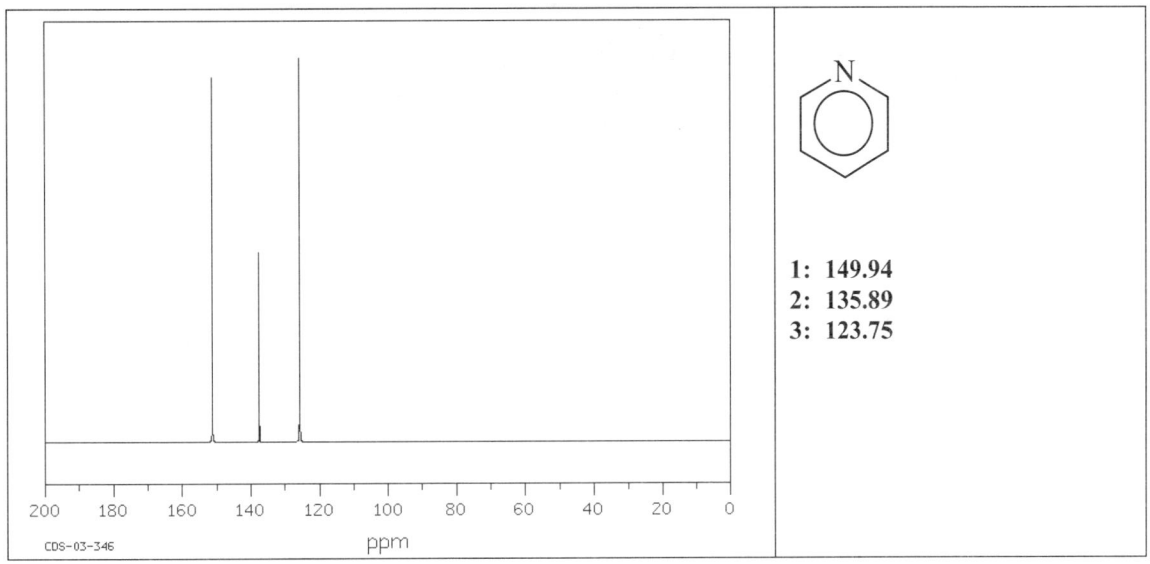

1: 149.94
2: 135.89
3: 123.75

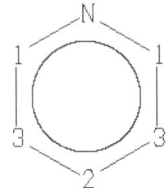

☐☐ **Assign the peaks in the following ¹³C NMR spectrum.**

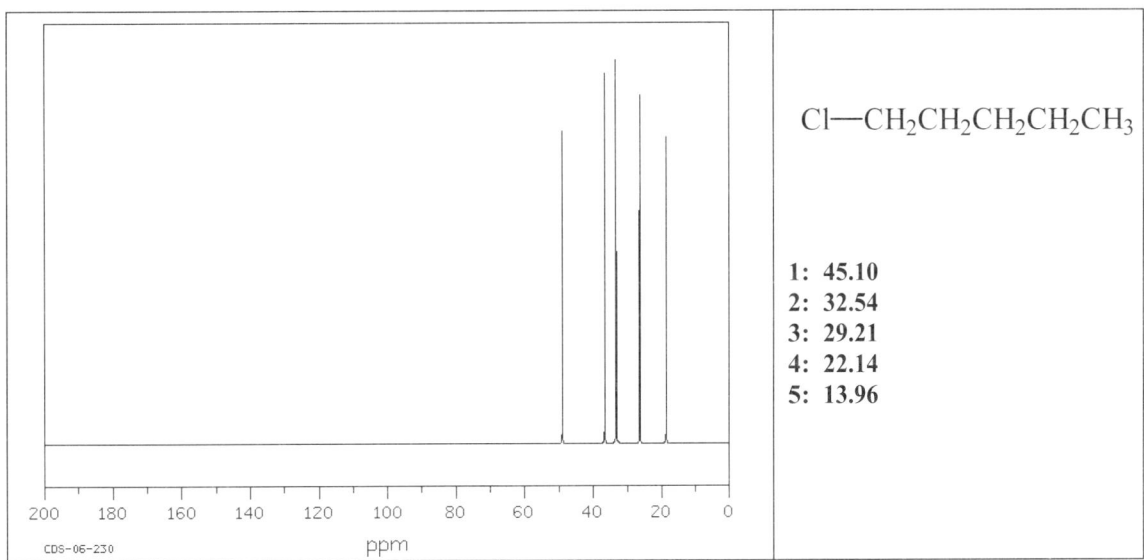

Cl—CH$_2$CH$_2$CH$_2$CH$_2$CH$_3$

1: 45.10
2: 32.54
3: 29.21
4: 22.14
5: 13.96

Cl——1——2——3——4——5

❏❏ **The signal for the proton on the Oxygen of an Alcohol is usually not split, no matter how many protons are on the adjacent Carbon. Why is this?**

Under most conditions, the proton on an Alcohol is constantly being exchanged between other Alcohol molecules. This exchange is very rapid compared to the NMR time scale and effectively decouples the Alcohol proton from all other protons in the molecule.

❏❏ **Under what conditions might the splitting of an Alcohol proton's signal be observed?**

The exchange is catalyzed by traces of acid, base, or other impurities. To see splitting the sample would need to be extremely pure.

❏❏ **When examining the ^{1}H-NMR spectrum of an Alcohol, how can you confirm that a particular signal is the result of the OH group?**

Add D_2O to the sample and shake it. If the signal disappears, then that proves that the proton was exchangeable and belongs to the OH group.

❏❏ **At room temperature, the proton NMR of Cyclohexane shows a single sharp peak. As the temperature is lowered, this peak broadens, and eventually splits into two peaks. How is this possible?**

If you consider the chair conformation of Cyclohexane, there are clearly two different types of Hydrogens: axial and equatorial. These two types of protons have different chemical environments and you would think they would have different chemical shifts. However, at room temperature cyclohexane is rapidly interconverting between the two chair forms. This interconversion swaps protons between the axial and equatorial positions. Since this occurs at a rate faster than the NMR time scale, at room temperature one averages signal is seen for all of the protons. At lower temperatures, the rate of interconversion slows and at low enough temperatures almost stops. Under these conditions the two different types of protons can be observed, and two signals are obtained in the NMR spectrum.

❏❏ **In IR, frequency is usually expressed in terms of wavenumbers. What are wavenumbers?**

Wavenumbers (symbolized by $\bar{v}$) are the number of cycles per centimeter, which may be calculated from the wavelength:

Wavenumbers (cm^{-1}) = $1 / \lambda$ (cm)

❏❏ **What wavelengths correspond to infrared light?**

The IR region of the spectrum goes from 7.8×10^{-7} m to 2.0×10^{-3} m. For analysis of organic molecules, only the region from 2.5×10^{-6} m to 2.5×10^{-5} m is usually used.

❏❏ **What are the usual IR scanning limits in wavenumbers?**

An IR spectrum is usually scanned from 4000 cm^{-1} to 400 cm^{-1}.

❏❏ **How does infrared light interact with molecules?**

It causes them to vibrate. The photons of infrared light have energies that roughly correspond to the energy differences between the vibrational energy levels of molecules (which are quantized).

❏❏ **What is necessary for a molecule to absorb infrared radiation?**

In order for a molecule to absorb IR radiation, there must be a bond or bonds in the molecule which are polar, and the vibration of these bonds must cause a change in the dipole moment of the molecule. The more polar the bonds are, the more intense the absorption. A single bond or group of bonds may have several different modes of vibration and may therefore absorb IR radiation of several different frequencies.

❏❏ **What are the simplest vibrational motions of a molecule?**

Stretching and bending. Even these simple motions have different modes. A methylene group, for example, has two different stretching modes and four different bending modes.

❏❏ **In what units are the wavelengths of infrared light usually expressed?**

In micrometers (μm). One micrometer equals 10^{-6} m.

❏❏ **On a typical IR spectrum, there are three scales. What are they?**

The horizontal axis is calibrated in both wavenumbers (cm^{-1}) and micrometers. Since micrometers and wavenumbers are inversely proportional, only one of these scales can be shown linearly. The vertical axis is usually labeled as % Transmittance, with 100 % being at the top of the spectrum (0 absorbance).

❏❏ **How are samples usually prepared for IR analysis?**

There are several ways of running an IR:

Neat: a liquid sample is spread as a thin film between two Sodium Chloride plates. Sodium Chloride is transparent to IR, so it does not obscure any part of the spectrum.

Solution: A solid or liquid sample is dissolved in a solvent that does not have strong IR Absorption in the region of interest. Typical solvents are CCl_4, $CHCl_3$, CH_2Cl_2 and CS_2. The sample is then placed in an IR solvent cell. A reference cell filled with pure solvent is often used in an attempt to subtract the IR signal of the solvent.

Nujol Mull: Solids are ground to a small particle size, mixed and ground with paraffin oil (Nujol), and spread in a thin layer between two sodium chloride plates. The Nujol provides a support material for the solid, but also has several intense IR absorptions. To get a complete spectrum it is sometimes necessary to run a second mull with another substance such as Fluorolube, which absorbs in different regions than Nujol.

KBr Disk (Pellet): A solid sample (1-2 mg) is ground with about 300 mg of KBr and then compressed at 14,000 - 16,000 lb/sq.in. This gives a transparent, glass-like disk, which may then be placed in the IR instrument. KBr is transparent to IR, so it does not obscure any part of the spectrum.

Gas Cell: Gases are analyzed using a special gas cell, which can be evacuated and then filled with a pure sample of a gas. A gas cell is much larger than a solution cell, because it must provide a much longer path length due to the low concentration of gases at atmospheric pressure. The windows of a gas cell are usually made of Sodium Chloride so as to be IR transparent.

❏❏ **Why do different functional groups absorb at different frequencies in the IR?**

A bond between two atoms may be conceptualized as a spring connecting two weights. The frequency of oscillation will depend on the masses of the two connected atoms and the strength of the bond. Since each type of bond will have different values for each of these three variables, each type will vibrate (and absorb) at a different frequency.

❏❏ **Is it possible to calculate the frequency of absorption for a particular type of bond?**

Yes. The stretching vibration of two masses connected by a spring may be described by Hooke's Law. The numerical relationship that describes two atoms connected by a bond is:

$$\bar{v} = \frac{1}{2\pi c} \sqrt{\frac{f}{(M_a M_b)/(M_a + M_b)}}$$

Where $\bar{v}$ is the vibrational frequency in cm^{-1}, c is the speed of light in cm/sec, f is the force constant of the bond (related to bond strength) in dynes/cm, and M_a and M_b are the masses of the two connected atoms in grams.

❏❏ **What would be the expected stretching frequency for a C-C triple bond? The force constant is equal to about 15×10^5 dynes/cm, and the mass of a Carbon-12 atom is 19.8×10^{-24} g.**

$$\bar{v} = \frac{1}{2\pi c} \sqrt{\frac{f}{(M_a M_b)/(M_a + M_b)}}$$

$$= \frac{1}{2 \times 3.142 \times 3 \times 10^{10}} \sqrt{\frac{15 \times 10^5}{(19.8 \times 10^{-24}) \times (19.8 \times 10^{-24}) / [(19.8 \times 10^{-24}) + (19.8 \times 10^{-24})]}}$$

$$= 2064 \text{ cm}^{-1}$$

❏❏ **The calculated frequency of the C-H stretching vibration is 3040 cm^{-1}, but the observed frequencies are between 2960 and 2850 cm^{-1}. Why is there such a large difference?**

The calculation using Hooke's law is an approximation, and does not take the environment of the bond into account. In addition, the force constants used for most bonds are average values measured using several different compounds. Considering the simplifying assumptions made, the calculation really isn't that far off.

❏❏ **An IR spectrum is usually broken down into four regions for purposes of memorizing the positions of specific IR absorptions. These regions stretch from 4000 - 2500 cm^{-1}, 2500 to 2000 cm^{-1}, 2000 to 1500 cm^{-1}, and 1500 cm^{-1} and below. What bond types absorb in each of these regions?**

4000 - 2500 cm^{-1} : C-H, O-H, N-H
2500 to 2000 cm^{-1} : C≡C, C≡N
2000 to 1500 cm^{-1} : C=C, C=O, C=N, N=O
1500 cm^{-1} and below : C-C, C-O, C-N, C-X

❏❏ **By what name is the region of the IR spectrum between 1300 -900 cm^{-1} known?**

This region is known as the fingerprint region. A large number of complicated absorptions appear in this region, making it difficult to interpret.

❏❏ **For what purpose is the fingerprint region used?**

Because the signals in this region are usually so complicated, the absorptions in the fingerprint region can be used to positively identify a compound. If the fingerprint region of a known and an unknown sample are identical, they are almost definitely the same compounds.

❏❏ **How are the intensities of IR absorption bands usually described?**

As strong (s), medium (m), or weak (w).

❏❏ **What is a correlation table?**

A correlation table is a list of various functional groups along with the frequency ranges at which they absorb in the IR. It is used as an aid in the interpretation of IR spectra.

❏❏ **How would you assign the major peaks seen in the following IR spectrum of Pentane?**

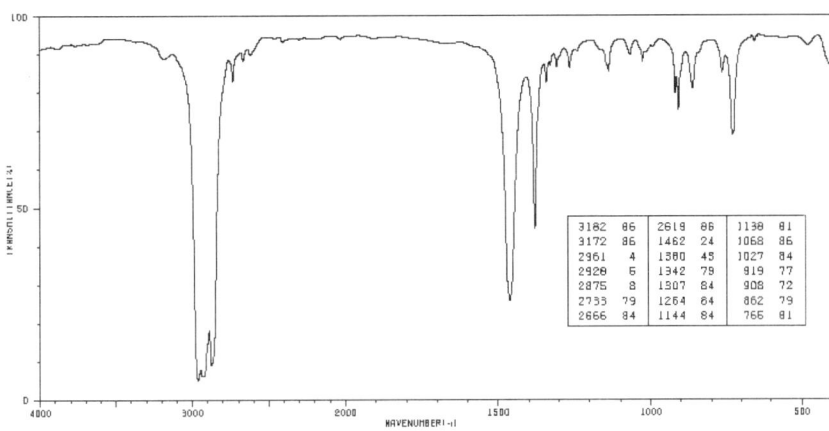

3000 - 2800 cm-1, C-H stretch; 1462 cm^{-1}, CH_2 bend; 1380 cm^{-1} CH_3 bend.

❏❏ **The characteristic C=C stretch of Alkenes (between 1600 and 1680 cm^{-1}) is often quite weak. What other signal readily identifies an Alkene?**

If the molecule has any vinylic protons, there will be an absorption between 3000 and 3100 cm^{-1}, just to the left of the aliphatic C-H signal.

❏❏ **What is the most characteristic IR signal of Alcohols?**

Alcohols show a very broad absorption band from 3200 to 3500 cm^{-1} corresponding to stretching of the O-H bond.

❏❏ **A broad absorption band very similar to that observed for Alcohols is often seen intermittently in the IR spectra of totally unrelated compounds. Why is this?**

Compounds are often contaminated with water, either from the atmosphere or from their preparation. The O-H bond of water appears in much the same place as that of Alcohols (3200-3500 cm^{-1}). This can be a particular problem for samples prepared as KBr disks because KBr is hygroscopic.

❏❏ **Carbonyls compounds are very easily recognized by the strong IR peak resulting from C=O bond stretching. Where does this signal appear in the IR?**

In the 1650 to 1800 cm^{-1} region.

❏❏ **How could you tell an aromatic hydrocarbon from an aliphatic hydrocarbon simply by looking at their IR spectra?**

An aliphatic hydrocarbon would have an absorption at 2850-2960 cm^{-1} corresponding to the C-H bonds. An aromatic hydrocarbon might have similar peaks, but assuming there were Hydrogens on the phenyl ring, it would also have an absorption at about 3030 cm^{-1}. A phenyl ring also shows two strong bands at 1600 and 1500 cm^{-1}.

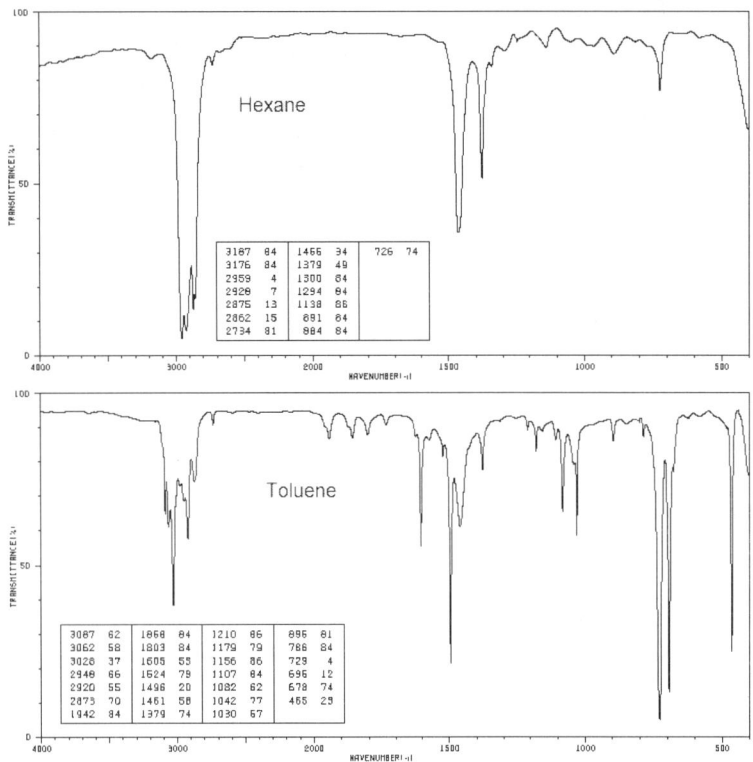

❏❏ **Amines have a characteristic absorption at 3300-3500 cm^{-1} in the IR due to the N-H bonds. How could you distinguish this absorption from the one seen for Alcohols?**

The absorption resulting from N-H stretching in Amines is much sharper and weaker than that usually seen for Alcohols.

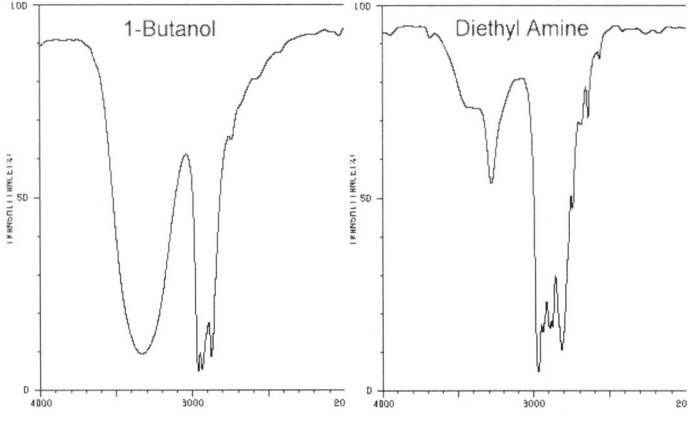

❏❏ **Where does the C-H bond of a terminal Alkyne show up in the IR?**

About 3300 cm^{-1}.

❏❏ **Organic compounds almost all contain C-C single bonds. Why are the IR signals from C-C bonds not very useful for structure elucidation?**

The C-C bending absorptions occur below 500 cm^{-1}, and therefore don't show up in IR spectra. The stretching vibrations are weak and appear over such a broad range (1200 - 800 cm^{-1}), that they are useless.

❏❏ **Shown below are the spectra of *o*-Xylene, *p*-Xylene, and Toluene. Which IR spectrum goes with which compound and why?**

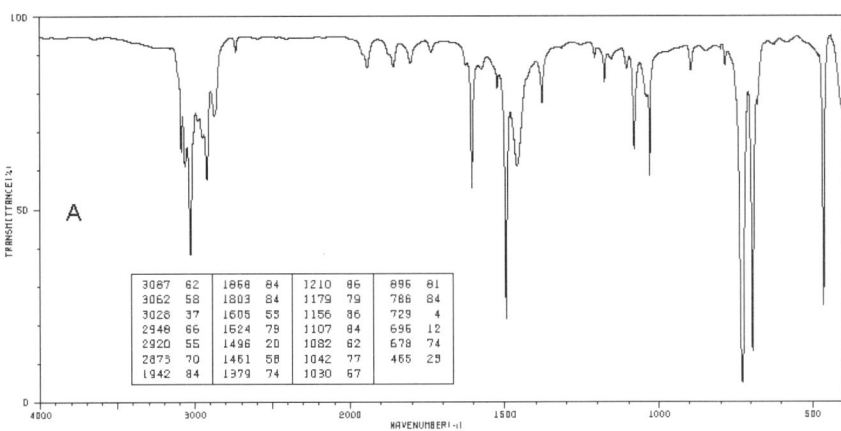

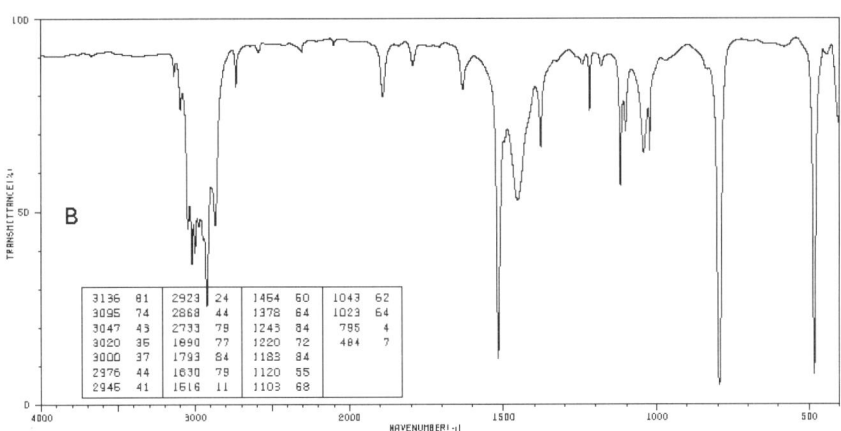

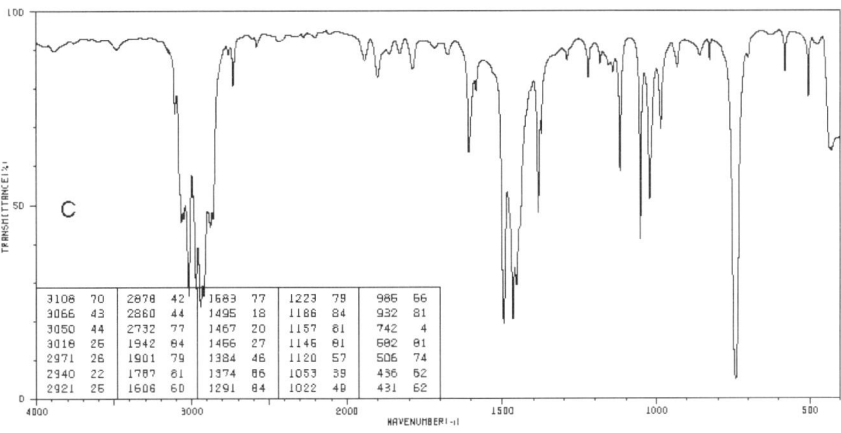

A) Toluene, B) *p*-Xylene, C) *o*-Xylene. Toluene is a monosubstituted benzene ring, and therefore should have a peaks in the IR at around 730-770 and 690-710 cm^{-1}. Only spectrum A has the two necessary peaks (at 729 and 696 cm^{-1}). *p*-Xylene, being a *p*-disubstituted benzene, should have a peak at around 790-840 cm^{-1}. Spectrum B has such a peak at 795 cm^{-1}. Finally, *o*-Xylene should show a peak at 735-770 cm^{-1}. This corresponds to spectrum C with a peak at 742 cm^{-1}. (Meta disubstituted benzenes, incidentally, have two absorption bands at 750-810 cm^{-1} and 680-730 cm^{-1})

❒❒ **What is the structure of a typical mass spectrometer?**

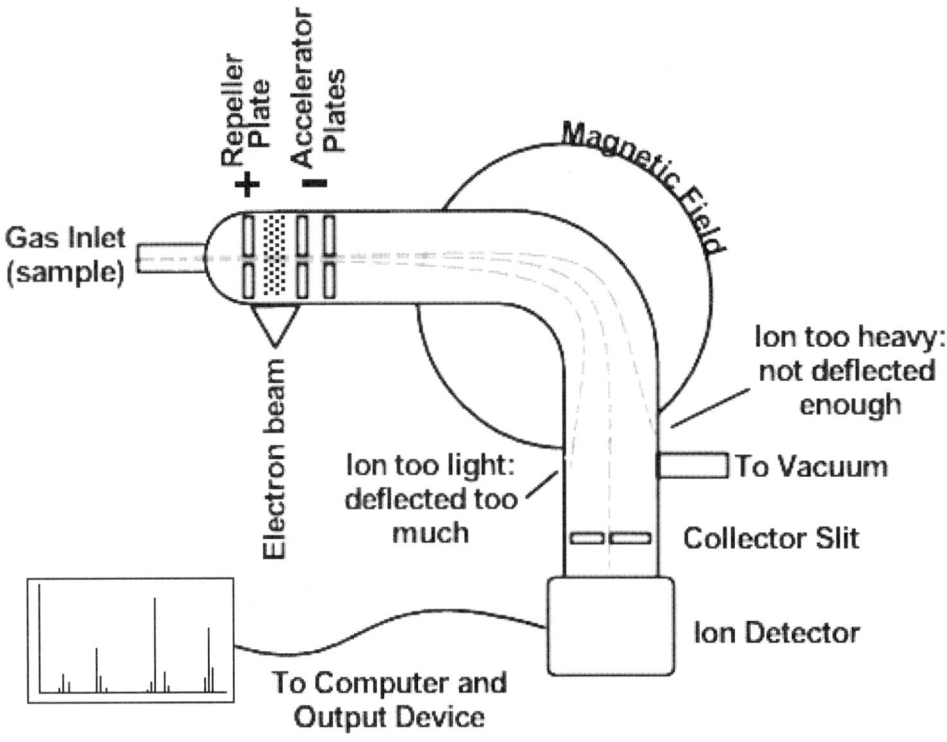

❒❒ **How does a mass spectrometer work?**

In the most common type of spectrometer, the sample is placed on a probe and inserted into a vacuum chamber. The probe is heated to vaporize the sample. A small gas inlet allows a stream of sample molecule to enter the rest of the instrument. Just inside there are three plates. The first is positively charged and the next two negatively charged. The sample passes through an opening in the first plate and then goes through a high energy electron beam. This beam knocks a single electron out of many of the molecules, leaving them as positively charged radical cations. At this point the particles are repelled by the first plate and attracted to - and through - the other two. Radical cations are fairly unstable, and these radical cations have lots of energy from the electron beam. They begin to break apart in predictable ways to form many different positive ions and neutral particles. As these particles enter the magnetic field, the neutral ones are unaffected and travel in a straight line, hitting the wall of the guide tube. The cations are deflected by the field, moving in a curve determined by their mass and charge (actually their mass to charge ratio, m/z.) At particular field strength, only particles with a certain mass to charge will pass through. Particle too light will be deflected too much, ones too heavy not enough. The right m/z will allow the particle to reach the detector where it is recorded. By varying the magnetic field strength it is possible to scan through all the masses and get a record of total number of particles (intensity) vs. m/z. since most particles have a +1 charge, what you really get is a plot of number of particles vs. mass.

❏❏ **What is a molecular ion?**

A molecular ion, symbolized by M^+, is the positive ion of the molecule under study, which has escaped fragmentation long enough to reach the detector. Some types of molecules always show the molecular ion in their spectra, some never do.

❏❏ **What is the difference between a low-resolution and a high-resolution mass spectrometer?**

A low resolution mass spectrometer distinguishes between ions that differ by whole mass units (by 1 amu). This type of spectrometer gives information about the nominal masses of ions. High-resolution mass spectrometers can distinguish masses that differ by as little as 0.0001 amu. This type of spectrometer provides information about the exact masses of ions.

❏❏ **Of what use is a high resolution mass spectrometer?**

Many ions with completely different molecular formuli have identical nominal masses, that is, masses calculated using whole number values for each of the isotopes ($^1H = 1$ amu, etc.). For example, N_2 and CO both have nominal masses of 28 amu. A low-resolution mass spectrometer could not distinguish between these two obviously different structures. A high resolution mass spectrometer can distinguish masses that differ by as little as .0001 amu. At that resolution a Nitrogen atom has a mass of 14.0031, a Carbon atom has a mass of 12.0000, and an Oxygen atom has a mass of 15.9949. This means that the precise mass of N_2 is 28.0062 amu and the precise mass of CO is 27.9949 amu. This is a more than large enough difference for a high resolution mass spectrometer to distinguish. Not only can a high-res mass spec tell apart ions whose masses are similar, if the molecular ion is present, its precise mass allows determination of the exact molecular formula of a compound. This is done by comparison of the mass with tables of precise masses of various combinations of atoms common in organic molecules.

❏❏ **What is the base peak in a mass spectrum?**

It is the peak with the highest relative intensity. The intensity of this peak is normalized to 100% and the intensities of all of the other peaks are reported relative to it.

❏❏ **What are the axes of a mass spectrum?**

The vertical axis is labeled Relative Intensity or % of Base Peak and the horizontal axis is labeled m/z (mass to charge ratio).

❏❏ **When a molecular ion cleaves, one of the products is a carbocation. What is the other product?**

Molecular ions are radical cations. When radical cations cleave, one fragment is a carbocation and the other fragment is a radical. Cleavage of a radical cation at any particular bond has two possible results (as shown below.) The radicals are not observed in mass spectra because they are not deflected by a magnetic field and so never reach the detector.

$$X^+ + Y^\bullet \longleftarrow [X-Y]^{\bullet +} \longrightarrow X^\bullet + Y^+$$

❏❏ **An electron beam is usually used to ionize molecules in a mass spectrometer. What is the usual energy of this beam?**

About 70 eV, strong enough to not only remove an electron, but also to impart a significant amount of energy to the resulting ion, causing it to fragment.

❏❏ **Which electron is usually lost when a molecule is struck by the electron beam?**

The electron that is lost is the one with the lowest ionization potential. Usually this is an electron from a π bond or from an unshared pair on a Nitrogen or Oxygen. Only when these are not present will electrons from σ bonds will be ejected.

❏❏ **What is the easiest way to understand the fragmentation patterns seen in mass spectra?**

Think of the patterns in terms of carbocation stability. Though radical cations fragment into a radical and a cation, the energy differences between the different types of radicals is very small. Therefore molecules are likely to fragment in such a way that they produce the most stable carbocation fragments rather than the most stable radicals.

❏❏ **Below is the mass spectrum of Chlorobenzene, which has a molecular ion of mass 112 when the mass is calculated using the most abundant isotopes. Yet there is a substantial peak above 112, at 114 m/z. How is this possible?**

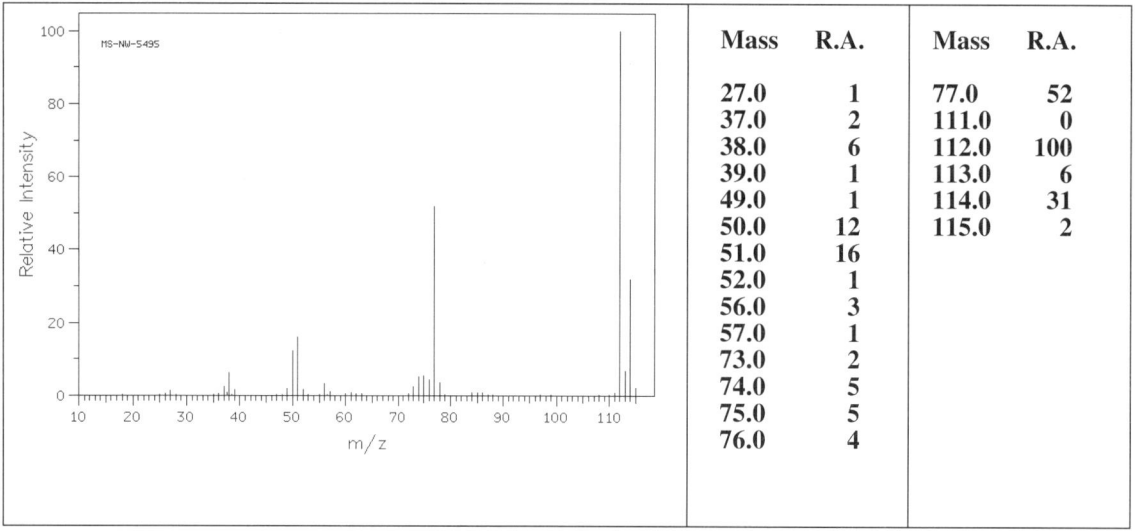

Mass	R.A.	Mass	R.A.
27.0	1	77.0	52
37.0	2	111.0	0
38.0	6	112.0	100
39.0	1	113.0	6
49.0	1	114.0	31
50.0	12	115.0	2
51.0	16		
52.0	1		
56.0	3		
57.0	1		
73.0	2		
74.0	5		
75.0	5		
76.0	4		

The mass of the molecular ion is calculated using the most abundant isotopes, i.e. Carbon = 12, Hydrogen = 1, and Chlorine = 35. Chlorine, however, has another isotope, ^{37}Cl, that is a significant portion of the naturally occurring Chlorine. The ratio of ^{35}Cl to ^{37}Cl in nature is 100:32.7. So there will be a significant number of molecules where the Chlorine is ^{37}Cl, and this gives rise to the observed M + 2 peak at 114. The ratio of the relative intensities of the M+ and M+2 peaks is 100:31, very close to the ideal ratio of 100: 32.7.

❏❏ **What other elements in an organic molecule will give rise to an M + 2 peak?**

Aside from Chlorine, the other two elements that are commonly found in organic molecules and that give M + 2 peaks are Bromine (^{79}Br: ^{81}Br = 100: 98.0) and Sulfur (^{32}S: ^{34}S = 100: 4.40.). Identifying M + 2 peaks in a mass spectrum and examining their intensities can help identify or confirm the presence of these elements.

❏❏ **Other elements common to organic molecules, such as Carbon and Hydrogen, have isotopes that are 1 amu higher in mass than the most abundant isotope. Would the M + 1 peaks arising from these isotopes have any analytical value?**

Not usually. The abundance of these isotopes that are 1 amu higher are very small, and so the M + 1 peaks would be very small relative to the molecular ion. Even if they were large enough to measure, they would be hard to measure accurately enough to make them of analytical value.

❏❏ **What happens if more than one Chlorine or Bromine atom is present in a molecule?**

One Bromine or Chlorine gives a distinct M + 2 peak in a mass spectrum. If there are two present, an M + 4 peak will also be observed. Three will give an M + 6, etc. The intensities of these peaks will decrease the farther they get from M +. Polychlorinated and polybrominated compounds may give some isotope peaks that are too weak to measure, but for molecules with fewer Halogens these peaks can help determine the number of Chlorines and Bromines present.

❑❑ **What is the Nitrogen Rule?**

If a compound has an even number of Nitrogen atoms, or no Nitrogen atoms, then its molecular ion will appear at an even mass value. If a compound has an odd number of Nitrogen atoms then its molecular ion will appear at an odd mass value. If there are an odd number of Nitrogens in a molecule, this rule can help establish their presence. If no Nitrogens are present, then this can allow an investigator to distinguish between the molecular ion and fragment peaks.

❑❑ **What is the Index of Hydrogen Deficiency and how is it determined?**

The Index of Hydrogen Deficiency is the number of degrees of unsaturation in a molecule, which is equal to the number of multiple bonds and rings. It is calculated from the molecular formula of a compound according to the equation:

$$\text{Index of Hydrogen Deficiency} = \#\text{Carbons} - \frac{\#\text{Hydrogens}}{2} - \frac{\#\text{Halogens}}{2} - \frac{\#\text{Nitrogens}}{2} + 1$$

❑❑ **What is the most recognizable feature of the mass spectra of straight chain Alkanes?**

A series of cations differing by 14 amu (one CH_2 unit). The molecular ion will usually be present, though it may be weak. The next lowest peak will result from loss of CH_2CH_3, since loss of CH_3 is never seen in these compounds..

❑❑ **The molecular ion is usually seen in mass spectra of straight chain Alkanes, but this is not always the case with branched Alkanes. Why is this?**

Branched Alkanes can form more stable secondary and tertiary carbocations, while straight chain Alkanes must form primary cations. Since these cations are more easily formed, extensive fragmentation is more likely in branched compounds, which destroys the molecular ion before it reaches the detector.

❑❑ **What is α-fission?**

In compounds with a heteroatom (such as Nitrogen or Oxygen), fragmentation of the molecular ion tends to occur at a position α to the heteroatom. This behavior is commonly seen in the spectra of Amines, Ethers, Aliphatic Alcohols, and carbonyl compounds.

❑❑ **What fragmentation patterns are commonly observed in Alcohols?**

Most common is loss of water to give an M - 18 peak and loss of an alkyl group from the Carbon bearing the -OH (α-fission).

❑❑ **Alkyl substituted Benzenes have a tendency to fragment at a bond β to the ring. Why is this?**

Cleavage of the radical cation at the β position gives a resonance stabilized benzylic cation, which likely rearranges to an aromatic tropyllium ion.

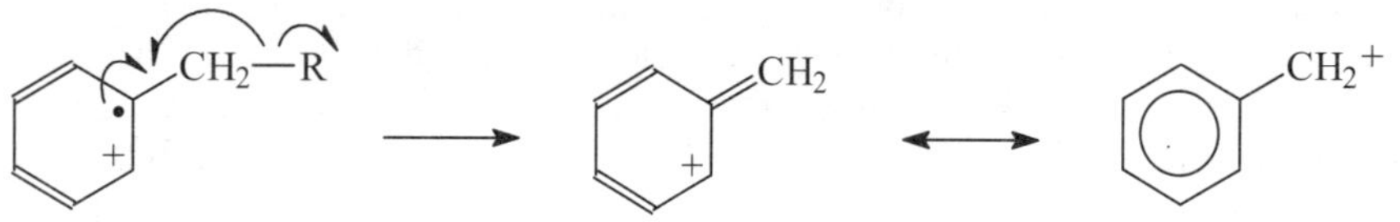

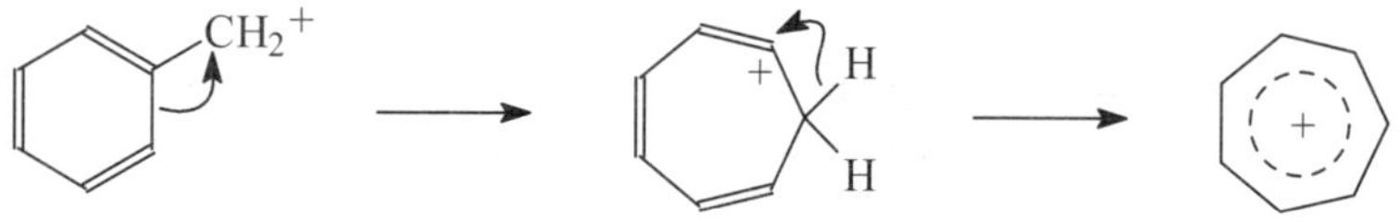

❏❏ What is the McLafferty rearrangement?

Aldehydes or Ketones of sufficient length can rearrange to give an Alkene and an enol radical cation. The rearrangement goes through a six-membered ring transition state.

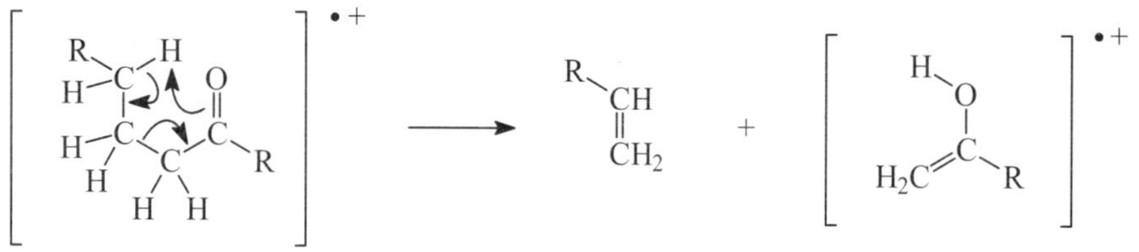

❏❏ Shown below is the mass spectrum of Acetone. Determine the structures of the major peaks and suggest a way in which they might form

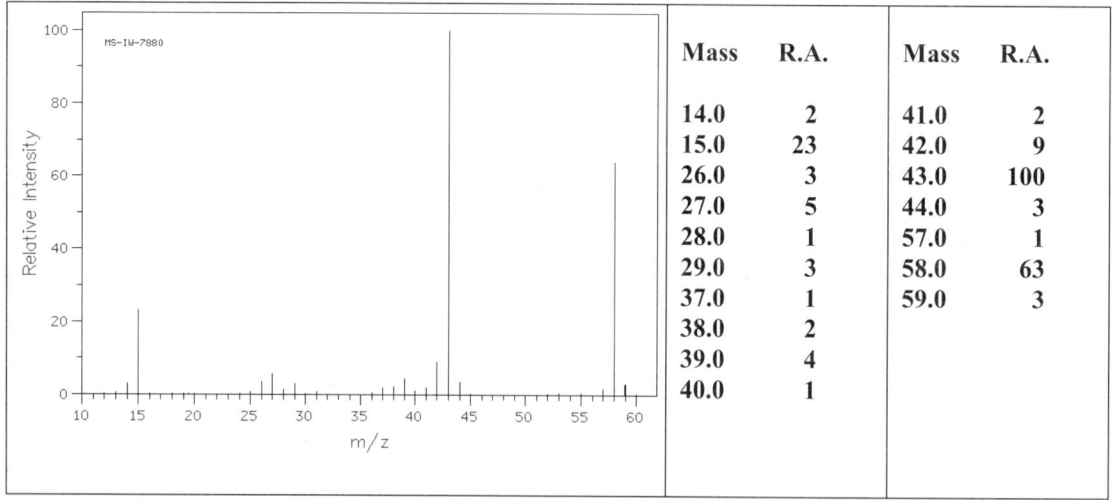

Mass	R.A.	Mass	R.A.
14.0	2	41.0	2
15.0	23	42.0	9
26.0	3	43.0	100
27.0	5	44.0	3
28.0	1	57.0	1
29.0	3	58.0	63
37.0	1	59.0	3
38.0	2		
39.0	4		
40.0	1		

Acetone has a mass of 58 amu, so the peak at that value is the molecular ion. The base peak at 43 amu is the result of loss of a methyl group (M - 15). This is an example of α-fission.

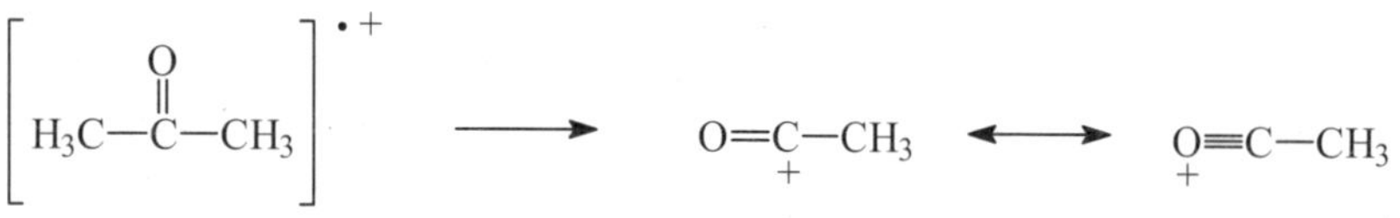

☐☐ **Shown below is the mass spectrum of Nonane. Determine the structures of the major peaks and suggest a way in which they might form**

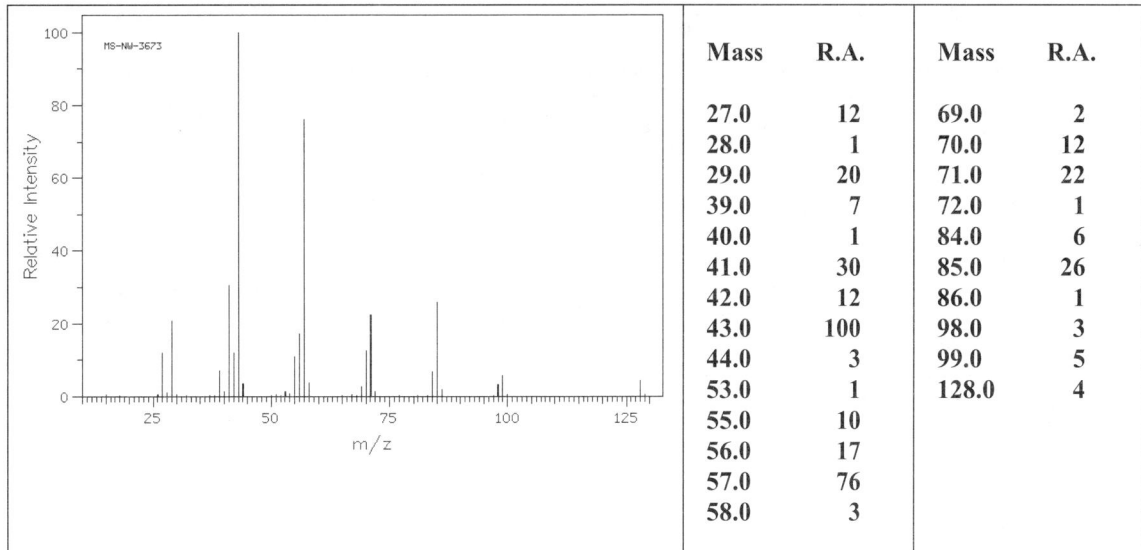

Mass	R.A.	Mass	R.A.
27.0	12	69.0	2
28.0	1	70.0	12
29.0	20	71.0	22
39.0	7	72.0	1
40.0	1	84.0	6
41.0	30	85.0	26
42.0	12	86.0	1
43.0	100	98.0	3
44.0	3	99.0	5
53.0	1	128.0	4
55.0	10		
56.0	17		
57.0	76		
58.0	3		

The molecular ion of nonane appears at 128 amu. The peak at 99 results from the loss of an ethyl group (128 - 29). The peaks at 85, 71, 56, 43, and 29 are 14 units apart and represent fragments with the structure C_nH_{2n+1}. These result from cleavage at each successive bond of the chain. The peaks grouped around these are C_nH_{2n} and C_nH_{2n-1} fragments.

$CH_3-CH_2-CH_2-CH_2-CH_2-CH_2-CH_2-CH_2-CH_3$ 128

$CH_2-CH_2-CH_2-CH_2-CH_2-CH_2-CH_3$ 99

$CH_2-CH_2-CH_2-CH_2-CH_2-CH_3$ 85

$CH_2-CH_2-CH_2-CH_2-CH_3$ 71

$CH_2-CH_2-CH_2-CH_3$ 57

$CH_2-CH_2-CH_3$ 43

CH_2-CH_3 29

☐☐ **How does ultraviolet and visible light interact with organic molecules?**

When a molecule absorbs UV-Vis radiation, this results in a transition of electrons from a low energy molecular orbital to a higher energy molecular orbital. The energy of this radiation usually corresponds to transitions within π molecular orbitals, especially conjugated π systems. The transition is from the π to the π^* orbital.

☐☐ **What are the axes of a UV-Vis spectrum?**

The vertical axis is labeled Absorbance and the horizontal axis is labeled Wavelength (usually expressed in nm).

☐☐ **What is the Beer-Lambert Law?**

When UV-Vis light travels through a solution, the absorption is proportional to the number of molecules that the light encounters. The numerical expression for this proportionality is $A = \varepsilon\, c\, l$.

Where A is the absorbance, ε is the molar absorptivity (also known as the molar extinction coefficient) expressed in liters per mole per cm, c is the concentration in moles per liter, and l is the path length in cm. This equation allows the quantitative determination of the exact concentration of an absorbing species if ε is known.

☐☐ **On what does the molar absorptivity depend?**

The molar absorptivity is a function only of the structure of the absorbing molecule.

☐☐ **Infrared spectra have very sharp absorption bands, but those in UV-Vis spectra are very broad. Why is this?**

The electronic transitions associated with the absorbance of UV-Vis radiation are usually accompanied by changes in the vibrational and/or rotational energy levels. These energy levels are very closely spaced and are superimposed on the electronic energy levels. It is impossible for an UV-Vis spectrometer to resolve these closely spaced levels, so broad absorption bands result.

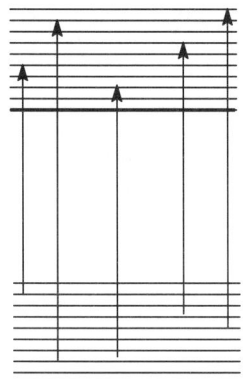

Excited State

Many slightly different transitions possible

Ground State

☐☐ **Which of the following molecules would you expect to absorb in the ultraviolet-visible region?**

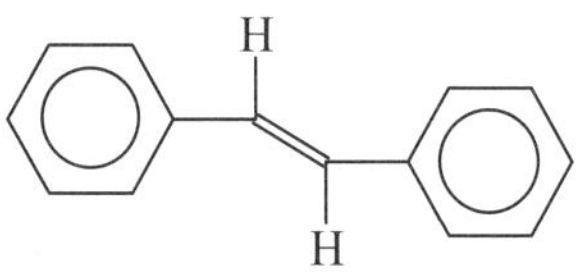

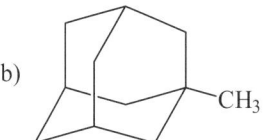

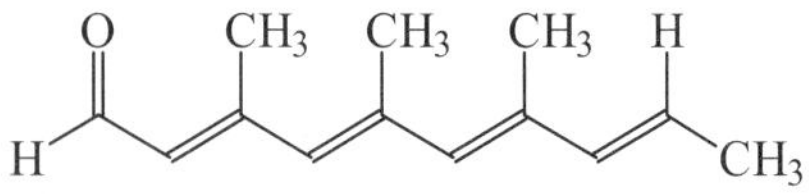

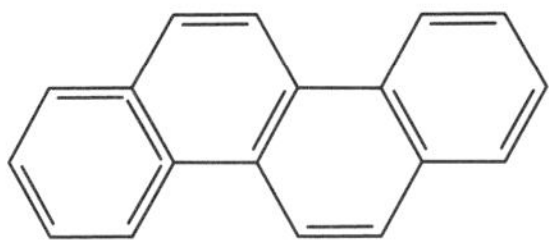

Compounds a, c, and d should all absorb in the UV-Vis region.

BIBLIOGRAPHY

Books/Articles/Websites

Brown, W. H. and Foote, C. S. *Organic Chemistry*, 2nd ed., Fort Worth: Saunders College Publishing; 1998.

Carey, F. A. *Organic Chemistry*, 3rd ed., New York: McGraw-Hill, 1996.

Fessenden, R. J. and Fessenden, J. S. *Organic Chemistry*, 6th ed., Pacific Grove: Brooks/Cole Publishing, 1998.

Lowry, T. H. and Richardson, K. S., *Mechanism and Theory in Organic Chemistry*, 3rd ed., New York: HarperCollins Publishers: 1987.

March, J. *Advanced Organic Chemistry*, 3rd ed., New York: John Wiley & Sons, 1985.

McMurry, J. *Organic Chemistry*, 3rd ed., Pacific Grove: Brooks/Cole Publishing, 1992.

SDBS: Integrated Spectral Data Base System for Organic Compounds, SDBSWeb: http://www.aist.go.jp/RIODB/SDBS/ (1999)

Silverstein, R. M., Bassler, G. C., and Morrill, T. C. *Spectrometric Identification of Organic Compounds*, 4th ed., New York: John Wiley & Sons, 1981.

Silverstein, R. M. and Webster, F. X. *Spectrometric Identification of Organic Compounds*, 6th ed., New York: John Wiley & Sons, 1998.

Williamson, K. L. *Macroscale and Microscale Organic Experiments*, Lexington: D. C. Heath and Company, 1989.